WETLAND HABITATS

A Practical Guide to Restoration and Management

NICK ROMANOWSKI

CSIRO

PUBLISHING

National Library of Australia Cataloguing-in-Publication entry

Romanowski, Nick, 1954–

Wetland habitats : a practical guide to restoration and management/Nick Romanowski.

9780643096462 (pbk.)

Includes index.
Bibliography.

Wetland management – Australia.
Wetland conservation – Australia.

333.9180994

Published by

CSIRO PUBLISHING
36 Gardiner Road, Clayton VIC 3168
Private Bag 10, Clayton South VIC 3169
Australia

Telephone: [+613] 9545 8555
Local call: 1300 788 000 (Australia only)
Fax: +61 3 9662 7555
Email: csiropublishing@csiro.au
Web site: www.publishing.csiro.au

Front cover: Mangrove jack, purple swamphen and freshwater mussels, set against a backdrop of water shield covering an inland billabong.

Back cover (clockwise from top left): Purple-spotted gudgeons; a flooded corner of a pasture rich in aquatic invertebrate species; green treefrog; rain falling on a permanent swamp on Stradbroke Island.

All photos by the author.

Set in Adobe Minion 11/13.5 and Adobe Helvetica Neue
Edited by Janet Walker
Cover and text design by James Kelly
Typeset by Desktop Concepts Pty Ltd, Melbourne
Printed by Ingram Lightning Source

CSIRO PUBLISHING publishes and distributes scientific, technical and health science books and journals from Australia to a worldwide audience and conducts these activities autonomously from the research activities of the Commonwealth Scientific and Industrial Research Organisation (CSIRO).

Contents

What this book is about

Over the past 30 years I have worked with wetland plants much of the time, since I founded the first specialist wetland nursery, Dragonfly Aquatics, in the 1980s, and have been researching plant and animal interactions since that time. Yet my background is in zoology, and I first began to study plants seriously by way of understanding their roles as habitat. Plants are only a part of what habitat is about, and are largely irrelevant to the habitat needs of many wetland animals other than in a general way as shelter, and for their role in improving water quality. For some others, they are just the vegetation that gets in the way of their hunting.

It has taken me around 30 years to put what I have learned about wetland biology, ecology and natural history into a practical form that can be used to understand what is happening in wetlands today, not just the natural processes but also the numerous impacts of what we like to call civilisation. Part of what has slowed the process down is the time it took to realise that quite a bit of what is popularly regarded as the basics of wetland management and restoration is a set of clichés and aphorisms rather than knowledge based on the ever-increasing literature and science of various aspects of wetland ecology.

Among other things I emphasise repeatedly is the need to shed our preconceived ideas of what looks to be an attractive habitat, and concentrate instead on what the animals themselves see, experience and are adapted for – their needs rather than our ideas. Many of the examples chosen show rich and diverse habitats that most people wouldn't look at twice, and there is a danger that in following clichéd management practices to satisfy our need for beauty, we may destroy some diverse and fascinating worlds while creating poorer habitats for a few of the most common, widespread and adaptable animals.

I have also tried to put a new and very different slant on the impact of birds on whole ecosystems, as the success of a wetland is all too often judged by the arrival of a few (or sometimes hundreds) of birds. These are the wild cards in wetland ecology, able to come and go so their presence is not necessarily a sign that a wetland is anything more than a way stop, and with powerful impacts on many other animals in ways that are not always obvious. Birds are an essential component in the ecology of most types of wetland, but it is a mistake to make management decisions purely on the desire to attract them in unnaturally high numbers, as is all too often done.

Finally, I have tried to show that what we call management is actually a type of repair work, trying to undo some of the effects of weeds, drainage, river de-snagging, channelisation, invading vermin and the diseases they are vectors for, and many other problems. If we hadn't arrived in this country wetlands would be managing themselves, and the starting point for all management must be an understanding of what the natural processes we are trying to restore originally were.

Acknowledgements

This book could not have been written without considerable help and encouragement from many people over a very long period of time, and it is impossible to remember the names of more than a fraction of the people who have helped me with research papers, ideas, suggestions, locations of particularly photogenic wetlands, and many other things. For this reason I have reduced the list of acknowledgements mainly to those people who have made specific contributions to this present book over the past two years, and apologise to the many people who have slipped through my memory.

I would particularly like acknowledge Ian Bayly and Bill Williams, as I was lucky enough to be doing zoology at Monash University at a time when they were leading a renaissance in the study of inland aquatic habitats, as well as Dave Morton from those days for a shared interest in the small things that creep in ephemeral pools.

My thanks for help and encouragement in many ways go to Paul Adam, Craig Allen, Gerry Allen, Helen Aston, Glenn Briggs, John Cann, Ian Clarke, John Dodson, Tim Entwisle, Michael Fenech, Bruce Hansen, Bruce Jackson and Marita Kennedy, Surrey Jacobs, Paul Koch, Tony Orchard, Tony Pickles, Tarmo Raadik, Mick Romanowski, Val Stajsic, Neville Walsh and Dave Wilson. I would also particularly like to thank Ted Hamilton at CSIRO Publishing for his timely encouragement to put together the proposal for this book, which might otherwise have been put off yet another decade!

Habitats

1

What is habitat?

There are many types of wetland, ranging from hidden ones far underground with their own unique fauna, through diverse and varied swamps, marshes, lakes and ponds, creeks and brooks, rivers and saltmarshes. All of them provide habitat, in one form or another, often for a great variety of living organisms, sometimes for just a handful of species able to thrive in extreme conditions.

The very word *wetland* suggests the idea of habitat in most people's minds, and is liberally used when talking about anything from the benefits of created wetlands, to restoration and management of degraded places. It is used so loosely that if common and adaptable birds such as sacred ibis or Pacific black duck are present, a wetland is deemed to be a success as habitat, though in these cases all it proves is that there is water present. In reality, most wetland animals (and many plants) need far more than a single 'habitat' for the long-term survival of the species.

Lifecycles and changing needs

Consider the short-finned eel (see Plate 16) of south-eastern Australia and New Zealand, which lives most of its life in fresh waters, feeding and fattening over decades until it has reached breeding age. Young female eels move inland from estuaries and colonise a wide array of wetlands from slow-moving streams and lakes, to farm dams. They are among the few native fishes which can climb past waterfalls, into areas where other fishes may be absent.

Their main needs for this stage of their life are a reliable source of food in the form of small animals including insect larvae and fishes, and a hiding place during daylight hours. The smaller males are more often found near the coast, and mature earlier than the much larger females, which may live more than 20 years in fresh water before they are large enough to spawn. Once a fish reaches maturity, it begins to reabsorb its digestive organs, because it will not feed again. These are converted into fat which will fuel its long swim to the breeding grounds, and into eggs or milt.

No-one knows exactly where short-finned eels breed, though it is believed to be hundreds of metres deep in the Coral Sea. The migrating adults probably travel along the ocean floor, but as they can't be caught in nets no-one is certain even of this. The one thing we do know is that none of them ever return to fresh waters, and that they probably die after spawning.

The young eels are leaf-shaped and transparent, and drift on ocean currents back towards southern Australia, feeding on smaller planktonic animals which drift with them. After up to three years adrift, they near the coast and change in shape into tiny but still transparent eels. Hiding in their thousands in estuaries, they develop dark pigmentation and grow a different type of teeth for a new life and diet upstream, before moving into fresh waters through spring and summer. Here we have a 'freshwater' fish which lives several different lives, spending most of its life as a major predator in inland waters (large females will even catch ducklings), as a non-feeding breeder on the ocean floor, and as a highly specialised planktonic drifter in the open sea. Its freshwater habitat is any place where it can successfully feed and grow over decades, but the species also needs two very different oceanic habitats as well to breed successfully, and reasonably pristine estuaries for the period of change to an adult form. Fresh waters are just one part of the mosaic of worlds inhabited by the short-finned eel, and its close relatives from elsewhere in Australia.

The common jollytail (see Plate 13) also lives much of its life in fresh waters, breeding in estuaries on a high spring tide. The eggs hatch with the next spring tide, and the young move out to sea for a few months before returning to fresher waters. Juveniles have been found hundreds of kilometres out to sea, which explains why genetically similar forms of this same fish are also native in New Zealand, Chile, Argentina and even the Falkland Islands in the southern Atlantic. This impressive range for a fresh water animal is even more remarkable as the wide-flung populations are genetically close, suggesting at least some ability to cross oceans while young.

Only a relatively small number of wetland animals have such extreme ranges and needs, yet a close look at the biology of many others shows comparably diverse needs at different stages of life. The common snake-neck turtle (see Plate 5) may live as an aquatic animal for most of its life, feeding and even hibernating underwater, and only emerging to sun itself or breed. Its eggs are laid on higher, relatively soft ground, and there is some evidence that individuals will shun places they have seen flooded, as this will drown the eggs.

The nests may be hundreds of metres from water, making the journey back a dangerous one for the young, which are so shy and elusive that they are rarely seen – not surprising, as they are taken by a wide range of predators from birds and larger fishes to water rats. Once they have reached a relatively large size there are few native predators which can harm them, and they may live many decades, perhaps even approaching a century in age.

Wetlands with reasonable numbers of adult turtles are often cited as good 'habitat' for this species, but there may be nowhere for them to breed, especially in urban areas. Even in less disturbed areas, predation on eggs can be high, and while goannas (a natural predator which has evolved with turtles over uncounted millions of years) account for a small proportion of losses, up to 90% of nests are now destroyed by the introduced fox. The combination of near-invisible young and long-lived adults can create a misleading idea of a healthy turtle population, yet they may be in an ecological dead end unnoticed by most people in their lifetime.

Flying animals, particularly birds, move freely between wetlands and may commute between vastly different habitats in a single day, or migrate over greater distances in response to drought or flood, predator pressure, competition and breeding needs. Many Australian species are excellent travellers, and will move from one end of the continent to the other if the wetlands in their vicinity dry up. Even such a ridiculously clumsy flier as the purple swamphen (see Plate 26) has been known to reach New Zealand when blown by strong winds.

The most widespread and common waterbirds are usually generalists, able to adapt to a range of climates and conditions. During their breeding period, wetland birds may use more than one habitat regularly. Ibis nest communally, often in trees and taller shrubs such as paperbarks, and may commute to feeding grounds elsewhere including flooded paddocks, estuaries, and in urban areas even to overflowing garbage bins. Ducks and swans usually breed close to where they will raise their young, while the versatile Pacific black duck may even nest on the edge of an apparently barren farm dam, if the grass surrounding it grows long enough to provide cover.

By contrast, most winged wetland insects are more reluctant travellers than birds, as flight is more of an effort for many of them, and is usually only resorted to if the waters they inhabit dry up. The best fliers among insects are dragonflies and damselflies, with some widespread species probably capable of travelling hundreds of kilometres to find suitable habitat and partners for breeding, so they may be as adaptable as the more common waterbirds. Yet the majority of dragonfly species are found within a limited geographic area, and even within that range are only found in specific places such as rocky streams through forest, near waterfalls, in peaty bogs, or even in mangrove swamps.

Many other examples of the diversity of habitat needs of various wetland animals, of variation in needs even between closely related species, and of changing needs over time and seasons are discussed throughout this book. A sweeping concept of habitat turns the word into a cliché that tells us nothing of value, and can even be used as a smokescreen to conceal the failings of a poorly managed or designed wetland. All goals in wetland management and restoration must be firmly anchored in what is known of the species-specific biology of any plant, animal or community we aim to maintain habitat for.

Perception of habitat

The most abused idea of wetland habitat is in the eye of the beholder – humans are prone to regard a wetland that looks good as one that *should* be a desirable residence for the species it has been managed or created for. Usually, an abundance of plants arranged prettily and a few species of birds (however common and indifferent to whether they are living in an estuary, river, lake or farm dam) is enough to confirm this impression. Humans value the presence of vertebrates (fishes, amphibians, reptiles, birds and mammals) more highly as evidence of a quality environment, regardless of whether these are just transients, than the often much richer array of invertebrate life present.

Water is an important part of the wetland aesthetic, so dry wetlands are regarded as less desirable, even though many aquatic animals are not only well adapted to regular, seasonal drying out, but may even depend upon it to stay one step ahead of their predators. Through the eyes of most animals, or more likely their finely tuned senses of smell, taste or hearing, a wetland must be a very different place to what we perceive as visually attractive. In the case of invertebrates, eyes are often of little use for perception in any sense we understand, and may be absent altogether!

Many of us have an anthropomorphic attachment to birds, and assume that we share some feelings with them about what they should like, even though our common ancestors parted ways hundreds of millions of years ago. After more than three decades of keeping and breeding diverse domestic and wild birds, however, I have been forced to the conclusion that they are probably much like the dinosaurs they evolved from, and where their ideas and ours on what is attractive as habitat converge is just a matter of luck.

Pelicans regard tips as excellent feeding grounds, as long as there is water nearby to excrete indigestible plastic bags into. Herons are attracted to garden ponds as long as there are goldfish in them, while some egrets think the world of cattle droppings. Swans will thrive on murky, plant-free waters with a scum of blue-green algae on the surface, as long as there are extensive lawns and other greenery at hand for them to browse.

Yet there is a strong tendency to believe humans know better what other species would prefer if they have the choice, and to design or manage wetlands accordingly. The classic example of this was the island design supposedly attractive to (unspecified) waterbirds, copied almost unchanged from one book to another through the 1980s and 1990s, and still in vogue in some circles. This usually featured hollow logs, whether lying down or propped up, and a selection of reeds, sedges or rushes spread out a little so there was room for birds to move about between them. It is surprising how few such islands have actually been made, and those I have seen are usually shunned by birds of all kinds, suggesting that the qualities we find most aesthetic are not what the birds themselves are looking for.

Most wetland animals have a far more restricted idea of what their surrounds look like than birds do, and many of them are so small or have such poor sight that they must appraise their world through very different senses to ours. The teeming water fleas (see Plate 11) and copepods (see Plate 21) in an ephemeral pool are preoccupied only with the availability of still smaller food sources ranging from bacteria to single-celled algae, and 'deciding' when it is time to shift from their most prolific breeding mode to laying hard-shelled eggs designed to survive the dry season to come.

In some of the biologically richest and most intriguing pools of this kind, the smaller crustaceans share their world with relative giants such as shield and fairy shrimps (see Plate 20), which are large enough to attract predatory birds as water levels fall enough to expose them. It is intriguing that the most abundant and varied-looking populations of these are almost always in ditches and farm dams fertilised by droppings of cattle and sheep, and also kept clear of vegetation by them.

The most conspicuous discrepancies between what humans perceive as attractive habitat and the needs of some wetland animals come in the form of insect damage. To human eyes chewed leaves on the elegant sacred lotus of northern Australia are unsightly damage (see Plate 10), while to the moth caterpillars which do the chewing, the shrivelled lacework is just the remains of a good meal which also happens to be their home.

Interactions

Satisfactory habitat for any animal is not just a matter of the right water quality, shelter, food and a suitable place to breed, but is also affected by its interactions with other animals in the form of competitors and predators. Fishes are one of the best examples of this, and can reduce populations of some wetland invertebrates to such low numbers that they may be unable to maintain adequate genetic variability for long-term survival – this idea is discussed in Chapter 8.

Fishes are also sometimes treated as the single greatest threat to young tadpoles, and therefore to frog populations overall, even though frogs and fishes have co-existed in many places for millions of years. As a result, a superstition has grown up among the owners of so-called 'frog friendly' garden ponds that all predators (but particularly fishes) must be excluded so that every tadpole can turn into a frog. This is contrary to all biological logic, as the reason many frogs produce so many tadpoles is that in a stable environment, only one offspring is statistically likely to survive from each parent *over its entire lifespan.*

If conditions are changing and a new ecological opportunity presents itself, many more offspring may survive although a new equilibrium will soon be reached. In most situations, however, the great majority of tadpoles are unlikely to survive long enough to breed, and even if predatory fishes aren't present so every single tadpole can reach maturity, there simply won't be enough places available for them all to move to. Predation is a part of the natural selection process any frog which shares the same waters as fishes has already been through, as have thousands of generations before that, so human intervention to attempt a new balance of species is unlikely to make a real difference in the long term.

Birds can make even more dramatic impacts than fishes, not just as predators but even on water quality if they are present in large numbers. For example, a large nesting colony of ibises or herons produces considerable quantities of guano, reducing oxygen levels in the water below as well as decaying to form toxic ammonia and other by-products. In a small body of water, enough wastes will build up to make the water uninhabitable during the breeding season except for a few invertebrates such as aquatic worms (see Plate 31).

The interactions of predator and prey also affect which other animals will thrive in a wetland. Predatory birds can reduce the numbers of predatory fishes, so that a greater diversity of invertebrates can thrive. In urban wetlands where bird numbers are artificially high, encouraged by bird lovers with an abundant supply of stale white bread, there are often enough ducks to keep fish numbers so low that a permanent population boom of water fleas and other crustaceans results.

Some invertebrates are also effective predators in a miniature way, including backswimmer bugs (see Plate 30) which often appear even in small and isolated pools, where they may eliminate mosquito larvae (see Plate 30) as well as feeding on other tiny species. In turn, backswimmers may be eradicated by the predatory larvae of dragonflies (see Plate 25), known as mudeyes in Australia. Mudeyes are significant predators on almost every other aquatic animal that isn't too large for them to handle, and the larger ones will devour fish and tadpoles several centimetres long.

2

Islands of water

Wetlands are as variable as any terrestrial ecosystems, differing not only in water quality but also in their soils, availability of oxygen, frequency of drought and flood, surrounding vegetation, and an ever-changing mosaic of interactions between all of these things and the plants and animals themselves. These factors determine whether there will be plants, the types of plant communities that will form, and in turn the presence or absence of plants also influences the types of animal habitat available.

Before looking at the many types of wetland, it is useful to understand the nature of the water which passes through them, as this has a direct impact on the health of many animals that even the best-laid plans for habitat will never be able to change.

Hardness and salinity

As water flows from higher ground to the sea or to saline lakes, whether over the surface or under the ground, it picks up anything soluble including organic acids, minerals and salts. Even when it first falls as rain it has already picked up a little carbon dioxide from the air to form traces of carbonic acid. The study of the movement of water and associated changes in its quality is called hydrology, and although this need not be considered in any detail here it is useful to understand the relationship of salinity and hardness.

Salinity is the measure of the amount of sodium ions in the water, with common salt usually the main component. Hardness is mainly a measure of calcium- and magnesium-based salts, though hard or saline waters usually have a mix of all of these types of chemical present. A general idea of how saline or how hard water is can be had by measuring the amount of the particular salts present in parts per thousand (ppt), and although this measure is little used by biologists these days it does give an image of just what is actually dissolved, and in what approximate quantities. Seawater is around 35 ppt, with roughly 35 grams of various salts present (mostly common salt) for every 1000 grams of water.

At lower concentrations, 1 ppt of total dissolved salts is unlikely to affect many animals apart from a few fishes and frogs which have evolved in really fresh waters, and for a very long time. By 3 ppt to 4 ppt many more animals and plants will be affected, though not necessarily to any worrying degree, but by around 6 to 8 ppt many plants will be weakening and will stop flowering, while frogs and some fishes will have disappeared. Other fishes, especially from inland waters, are much less affected as they have always lived in saline, alkaline environments, as have many freshwater crayfish – for example, yabbies and marron will survive concentrations up to a third that of seawater, though not necessarily thrive in the long term.

Hidden waters

There is far more water hidden underground than in the more readily visible surface wetlands, although most of it is held in soils and sand rather than forming pools, and as it moves through rock or gravels it usually becomes increasingly saline and often alkaline (caustic) as well. In extreme cases such as the Great Artesian Basin which underlies much of the drier parts of Australia this water can be undrinkable, though many fishes from desert gobies to the rare and very localised red-finned blue-eye thrive in such waters where they surface.

In limestone areas such as the Cape Range of tropical Western Australia (see Plate 2), even the slight acidity of the occasional rainfall is enough to gradually dissolve the stone, carving out caves and sinkholes over millennia. These hidden but often connected waters have developed their own unique fauna, including blind cave eels, gudgeons and shrimps. Little is known of these cryptic fishes as they are only occasionally seen in sinkholes where the subterranean waters are open to daylight.

In mountainous country where the water is still relatively pure, water also flows mostly underground and out of sight except after heavy rains, sometimes surfacing as springs, though these may also be drawn up from greater depths by capillary action. Water seeping down through fern gullies, however slowly, eventually leads to and forms the small, often clear streams which can be important feeders for larger rivers in times of drought.

Even while flowing underground this water carries enough oxygen that rich communities of ferns can form, though these would drown in more stagnant conditions, and further downstream ferns remain dominant along the banks of shaded permanent waters and other wet areas (see Plate 12). They are most abundant where water is relatively soft and lacks minerals, and light levels determine which species are present. Treeferns and water ferns dominate in shaded places, while on brightly lit wet sites masses of coral and fan ferns may scramble among paperbarks. The paperbark canopy can be so dense that possums remain active even on bright summer days, while at ground level the chimney-like burrows of land yabbies are common.

Damplands and heathlands are another hidden reservoir that helps to maintain a reasonably uniform stream flow, and where these cover large areas, the slow movement of water through them builds up large amounts of organic matter in the form of peat. This is basically plant matter which has only partly decayed while submerged, and is usually acid except where flooded by more saline waters. Moist peat can absorb an impressive volume of water, releasing it gradually into streams and wetlands, but if it dries out completely it becomes water-repellent and can burn for years underground if it catches alight.

Damplands and heathlands aren't usually recognised as wetlands because much of their water is below the soil, sinking deep during drier seasons and rising to saturate the surface during wetter times. The level of this underground water is called the water table, and where it rises above the soil level during the wet season to fill all depressions with water and leaving even the higher ground close to sodden, it may be called sumpland.

Damplands are often acid and poor in nutrients, yet may support a diverse range of flowering plants including teatrees, paperbarks, heaths, a variety of sedges from tiny creeping species to giant sawsedges, and other grass-like plants from cordrushes to yellow-eyes. The shortage of some nutrients here (especially nitrogen, phosphorus and potassium) has been the cause of much specialisation among plants, including so-called carnivorous plants which obtain their nutrient needs by trapping insects.

These low-nutrient environments may not attract many larger animals apart from some frogs and smaller mammals, but they provide habitat for some highly specialised invertebrates. The extensive buttongrass plains of southern Tasmania (see Plate 2) are eventually shaded out by encroaching teatree except where fires are fairly frequent, and it is often assumed that it was aboriginals who kept them open through regular burning. Yet a species of the ancient and primitive crustacean *Allanaspides* lives only in small pools and marsh yabby burrows in these plains, suggesting that natural fires were frequent in these wetlands long before humans arrived, and they are also an important seasonal habitat for the endangered orange-bellied parrot.

Seasonal wetlands

Ephemeral wetlands often fill directly from rain runoff, or from floodwaters and nearby streams, lasting for only a few weeks or months before drying out through evaporation (see Plates 3 and 21). In southern Australia, these usually begin to fill with autumn rains, vanishing some time after the last spring rains, while further north the well-defined wet season with its associated flooding is at the time of the southern summer.

The plants in ephemeral wetlands may die back to corms, rootstocks or leave a buried reserve of seed until better conditions return, while some of the animals move elsewhere, and many others produce drought-tolerant eggs or cysts. Many frogs are specialist breeders in ephemeral pools, and huge numbers may appear within a few hours of a heavy downpour. As the pools dry out in more arid areas, some specialised species of frog will burrow into the ground where they may wait passively for years until the next rains.

While such conditions seem harsh to us, they have great advantages for suitably adapted organisms. Ephemeral wetlands with extensive areas of open water, however shallow, may swarm with life for a time, and are free from many types of predator which don't have the adaptations needed to survive long dry spells. These open waters will eventually draw a range of predators including larger insects, in turn fed upon by birds ranging from herons to terns, all able to fly elsewhere as the water and their food supply dry up.

Grassy wetlands are a variation on more extreme ephemeral wetlands, retaining enough moisture for the survival of extensive areas of plants between wet periods. Covered in a mix of drought tolerant grasses and sedges, they support an underground fauna of tiny burrowing animals, and provide shelter and food for small mammals. Tussock-forming plants are especially useful as shelter since they tend to space themselves out, with the natural passageways between the tussock bases protected from flying predators by wiry, often sharp-edged foliage above.

Marshes and swamps

Extensive areas of still, reasonably permanent wetlands are called marshes or swamps (see Plate 3), and these varied and often species-rich wetlands can spread over great distances in some seasons. Though their water levels may rise and fall to some degree between the wet and the dry times, usually some water remains near the surface or in a few deeper pools except in the driest years, acting as a refuge for aquatic animals which can't fly or walk away.

There are no simple generalisations that can be made about marshes and swamps; even two adjacent swamps can be dramatically different in the plants and animals which dominate them. In part, this may be because the first two or three plants to establish in a swamp are often able to prevent later arrivals from establishing themselves properly, as many wetland plants actively deter competitors with a complex chemical arsenal. This is discussed in more detail in a companion volume, *Planting Wetlands and Dams* (Romanowski 2009).

In turn, many animal species may only be found where certain plants are already established, although this is better documented for insects rather than for larger animals. Numerous frog species breed in swamps and marshes in preference to any other habitat, as do many fishes. Where there are extensive stands of sedges, these attract nesting birds, while more open areas offer opportunities for cormorants, pelicans and herons as feeding grounds (see Plate 8).

Marshes and swamps are among the hardest hit wetlands in post-1788 Australia, as they have been drained for agriculture, grazed by livestock, poisoned to control mosquitoes, and mangled by the introduced feral pig and buffalo. Each swamp drained may take with it a range of species not found anywhere else in the vicinity, and the distances between the remaining wetlands also become greater, so fewer animals can move between them. Isolated swamps and marshes are effectively islands of water in a sea of land, and their increasing isolation will further endanger species already on the brink in increasingly arid areas.

Rivers, creeks and associated waters

Rivers, creeks and streams need no definition, although they can vary considerably in type from cold, fast-flowing mountain brooks to sluggish, warm and usually muddy inland rivers (see Plates 1, 4, 5, 15 and 18). These are more like bridges or roads than islands, though many indigenous wetland animals and plants are too specialised for life in particular types of marsh or swamp to be able to commute via moving waters. By contrast, the most successful introduced aquatic vermin including carp, plague minnows and tilapia spread quickly and effectively along streams.

Waterfalls are a particularly specialised stream habitat, acting as a barrier to upstream movement for some animals (and in turn offering opportunities for those able to climb or fly) to streams with a much more restricted range of indigenous predators. The deep, well-aerated pools at their base, constant humidity, and an almost complete absence of grazing animals on the steeper slopes allows unusual communities of plants which are found nowhere else to evolve, from mosses (see Plate 2) to moisture-loving grevilleas.

Rivers flowing through floodplains (particularly near the coast, and through much of the Murray–Darling system) tend to meander as they cut their paths across. These plains of silt are deposited during periods of flood and may spread widely, and the various depressions carved through them form swamps and pools after floods (see Plates 14 and 32). The most familiar name for these in Australia is billabong, a river bend which has been isolated from the main river flow when a new channel is cut, and gradually silts up at the openings leading to the main river channel.

Billabongs and other floodplain swamps are fascinating worlds in their own right, both a part of the river and separate from it. Their fauna and flora can be very rich, and may include species which aren't often seen in other types of wetland.

With the increasing control and regulation of many rivers, particularly the inland ones, new billabongs are now rarely formed on floodplains where they would once have appeared after almost any heavy flooding. At the same time, the disruption of *natural* flooding is reducing the species diversity of the old billabongs.

Lakes

Lakes also vary considerably as these are formed in many ways - tarns scoured out by glaciers in the high country, volcanic craters where fresh water from one can spill into a lower salt lake nearby, valleys dammed by lava flows or landslides (see Plate 16), hollows behind sand-dunes which accumulate organic matter until they hold water. Their waters vary from nearly as pure as distilled water, through acid and peaty, clear or soupy with microscopic life, and brackish to more salty than the sea.

Most of our lakes are less than 10 000 years old, formed since the end of the last Ice Age, and with the onset of the drier conditions in southern Australia in the past decade, many of the shallower ones are disappearing as rainfall fails to keep up with evaporation. Some are much older, reappearing whenever climate and rainfall are suitable, including some ephemeral lakes of central Australia with a pedigree going back millions of years.

The larger a lake is, the less diverse the vegetation around its fringes will be, sometimes limited to just one or two of the most wind- and wave-tolerant sedges or reeds. Lake-dwelling animals may also need to be tolerant of other extremes; for example, in deeper lakes the upper and lower layers may have quite different conditions for most of the time, only mixing at the two times of the year when the surface and bottom layers equalise in temperature.

Worlds in a teacup

At another extreme, some wetlands hold only a few litres of water, yet may support a surprising variety of life-forms. The mosquito-wriggler-filled ice-cream container under the front verandah is an all too familiar example, and is a more reliable breeding ground for mosquitoes in suburban areas than most wetlands. Even temporary rainwater pools at the grassy fringes of playing fields and cow pastures develop their own fauna if the water stands long enough to decay grass and other organic matter, providing food for abundant bacteria, and later other and slightly larger animals still barely visible to the naked eye (see Plate 21).

On the surface, springtails cluster in drifting clumps, looking like a grey, waxy mould until you look closely to see hundreds of minute, waxy, almost formless animals. Incredibly resistant to dessication, springtails will appear within hours of a pool refilling, and are an example of how little we know about most of the smallest inhabitants of wetlands. Tree hollows haven't been studied in Australia much either, but are known to have a distinct and specialised insect flora in the Americas, some burrowing in the sediments at the bottom while others are free-swimming or are adapted to creeping along the sides of the hollow.

Saline wetlands

Saltmarshes have a very distinctive flora much of which is shared between inland areas, and the more open areas behind mangrove belts on the coast (see Plate 6). At first glance, they may seem rather bleak because of the absence of trees, which can't establish because during periods of low water flow salinity levels are beyond the tolerances of most plants. A closer look will reveal some fascinating plants adapted to the extreme conditions prevailing, often arranged in zones which reflect their individual tolerances to flooding with fresh water or salt, and to varying degrees of drought at some seasons.

The animals in saltmarshes are mostly inconspicuous other than migratory birds, for which these may be essential feeding grounds as they pass through, and are a mix of specialised saltmarsh species and visitors from both fresh and marine waters. Many of them are mobile species such as snails and crustaceans which can shift away from conditions not to their liking, or burrow to avoid hot sun and dry conditions in summer. More terrestrial species including spiders must be able to reach the air during flood tides, or carry air pockets with them under water.

By most definitions, wetlands end somewhere along the sea's edge, and where the transition is fairly gradual there will often be trees or shrubs called mangroves (see Plate 7) – below their roots in the zone of lowest tides, the seagrasses begin. Mangroves are largely found in estuaries, where sea and fresh water mix; the sea pushes upstream at high tide, rivers and creeks push the saltwater back downstream as the tide falls. The mixing of fresh and salt water varies with flood and tide to give varied salinities at different times and places through the estuary, although in smooth conditions the less dense fresh waters may ride out to sea above the denser seawater below.

Saltmarshes are at their most varied and extensive in southern Australia, being replaced by an ever-increasing diversity of mangroves as we move further north, partly because these shade out the lower growing saltmarsh plants. Mangroves are not a single, natural group of related plants — with representatives from nearly 20 different botanical families including a palm, ferns, and relatives of eucalyptus, poinsettia and persimmon, they are more a guild of plants with specialisations for fluctuating but overall saline environments.

Mangroves and the estuaries in which they thrive are perhaps the most important feeding and nursery grounds for the young of diverse marine animals including the young of many commercially important marine fishes (see Plate 7). Nearly all Australian freshwater fishes are descended from marine ancestors as well, and although quite a number of these now live and breed entirely in fresh waters, many other species must return to estuaries to breed, and their young often have a marine stage in the lifecycle.

Boundaries

The difficulties of drawing clear boundaries between one type of wetland and another have already been noted, but wetlands are also often in a state of change, whether this is slow as in the gradual silting up of a backwater, or it happens overnight. It doesn't take long for a river in flood to cut through a meander to leave a backwater which will become a billabong, or for a storm to block the entrance to a brackish coastal lake, or tear it open so it is flushed out by fresh waters.
As wetlands are by definition the lowest point of ground through which water can flow, on low-lying ground even a slight elevation or depression of the land through minor earth movements can change the directions of streams and the boundaries of water catchments. Peaty, fresh water swamps behind sand dunes will inevitably become saltmarshes as they fill with sediment and the sea level rises.

Even the rivers which link many other types of wetlands can change dramatically over short distances. For example, the Barwon River was blocked by a landslide in the 1950s to form a lake seven kilometres upstream from where I live (see Plate 16). Originally 25 metres deep, this has now silted up to a degree where it has become a feeding and breeding place for a family of platypus.

Below that it resumes its original course as a clear mountain stream flowing among erosion-smoothed rocks, vanishing underground for around a kilometre during dry weather, then spreading out onto a series of narrow flood plains with numerous backwaters and small billabongs both up- and downstream. Further downstream it becomes increasingly saline as it progresses through farmlands and past salt lakes, then the city of Geelong, and opens into a great saltmarsh at Lake Connewarre about 80 kilometres from here before reaching the sea – yet even these changes over distance are trivial compared to those wrought over geological time.

3

Natural change

The harsher an environment, and the more extreme or prolonged the changes it experiences, the fewer species of plants and animals will be able to survive in it over the long term, but in favourable conditions (which include the absence of most potential competitors and predators) these few may appear in vast multitudes. This applies as much to wetlands as to the highest tidal zone of the sea, or a desert in bloom after a brief downpour.

Paradoxically, most wetland plants and animals are so well adapted to change that species diversity in (for example) a billabong may decrease if it remains permanently flooded. The refilling of a billabong after a dry spell will trigger seed germination of aquatic plants which will colonise areas where more terrestrial species grew for a while, until drowned by the rising waters. As the drowned plants decay they fertilise plankton blooms, providing an abundant food source for young animals such as tadpoles and fish fingerlings. In turn, many of these become food for passing waterbirds hunting over exposed mudflats (see Plate 28) as the water levels fall again during the next dry season.

More permanent waters are less common but are essential for animals which can't migrate or survive drying out, particularly fishes. In Australia, nearly all of these are predators, whether on microscopic or larger animals, and their presence changes the ecological balance where they are present. This is most obvious in northern waters where fishes can breed and disperse on a grand scale during the wet season (see Plate 15), most of their offspring ending up trapped in temporary waters, yet the relatively few survivors will remain as significant predators in almost every permanent pool that was linked by the floods that came before.

Change isn't something that the aquatic biota has passively adapted to – it is the driving force of evolution, selecting for those individuals of each species that are best able to breed or even just survive through fluctuating conditions. Only the most aggressive pairs of swans are likely to be able to defend nests with adequate foraging areas around them in times of drought. Only the most cold-tolerant subtropical fishes will survive a combination of extreme cold and low waters in more southern latitudes, to breed when conditions are on the mend, inching their way further south generation after generation as they adapt to new and more extreme environments.

The most dramatic changes in wetlands are over long periods of time, as they shift or silt up with land movement, become more saline after their drainage is blocked by earth movements, or disappear altogether as climate and sea levels fluctuate around them, setting the overall framework within all other types of change must be considered.

Change over time

Eighteen thousand years ago Australia was larger and much drier than now. Sydney Harbour was a pleasant, wooded valley, and you couldn't even hear distant surf on the plain which would become the Great Barrier Reef. Freshwater plants and animals moved fairly freely between northern Australia and New Guinea, which were joined into a single landmass called Sahul, and this is why so many species on these two big islands are closely related or even virtually identical today.

Most of the world's fresh water was locked away in the ice caps and glaciers of the northern hemisphere, but the ice was melting and over the following millennia sea levels rose perhaps 130 metres, before falling a little to the present level. Some droughts during this time span lasted centuries, and hiccups in the world's changing climate also caused occasional slips back to conditions a little more like the Ice Ages. All of what we humorously refer to as civilisation has happened in just the last 10 000 of those 18 000 years, including the development of agriculture, cities, genocide and untrammeled global tourism.

If I could step back through those 10 000 years, the world within a day's walk from the desk I am writing at would be a very different place, and the local rivers, streams and wetlands would be almost unrecognisable. The cool, fast-moving and perennial streams flowing south from the ranges I can see over the top of my computer screen were dry gullies, except after heavy rains, which is why the only fishes living in them now are those with a marine stage in their life cycle. Where those dry gullies crossed the future site of the Great Ocean Road, today one of Australia's iconic tourist attractions, the sea was just a distant sheen to the south-west, and a few centuries earlier you could still have walked from my front door to Tasmania.

A few true freshwater fishes survived in the Barwon River which passes a kilometre away, and may have been little more than a chain of ponds at times, but must also have carried some impressive floods. We know this because there were still active volcanoes a few kilometres to the north, and a lava flow from one of them blocked the Barwon River, diverting it west to form a huge body of fresh water. With an area of over 250 km^2, Lake Corangamite is still by far Australia's largest 'permanent' lake, but it was seven times larger then.

It is impossible to imagine how changes on this scale must have affected the smaller wetlands in the area and their biota, yet the causes of the changes which have taken place through all of that time until the present were natural. They were acting on a world largely undisturbed by humans, if you overlook the earlier extinction of much of Australia's megafauna, and the consequent ecological rearrangements which must have been set in motion.

All aquatic animals and plants which could not adapt to the changing conditions, or disperse to new homes as sea levels rose, simply died out. Some less adaptable or mobile species may have largely disappeared under rising seawaters, hanging on in small pockets on the mainland, and this may explain why some animals and plants are rare or very localised. Much of today's aquatic biota, however, is reasonably widespread, whether it walked, swam or flew to new homes, or seed, spores, cysts and eggs were blown or were spread by other agents.

Their survival and continued existence today reflects great adaptability in the face of dramatic changes caused by natural processes, albeit at a time when there was little competition from introduced weeds and vermin. That survival is also dependent on a range of strategies and abilities which enable them to live with regular or even intermittent changes, the most obvious and (on average) predictable of which are seasons.

Seasonal change

Australia is a huge island, stretching from southern areas with regular snow and frost to the tropics, yet all of these places have distinct seasonal changes typically associated with changes in rainfall. The animals and plants living in each area are not only fully adapted to these changes, but often use them as cues for breeding, flowering or seed production.

To the north, a relatively short but intense wet season creates flooding across extensive areas, allowing breeding animals to disperse into new parts of their range, but also acting as a death trap for those which end up in waters which will dry out completely before the next rains. The newly flooded lands are rich in decaying matter, fuelling explosive growth in diverse invertebrates such as tiny rotifers and crustaceans, ideal food for recently hatched fishes, and insects also abound in some waters and may provide flying feasts for those creatures which feed above.

In these areas plants tend to time their flowering and seed production for after the floods, as the freshly exposed areas have often been scoured of competing wetland vegetation, and terrestrial species have drowned and decayed away. Annual species especially benefit from the clean slate the receding floodwaters leave, and may flower and set copious amounts of seed in just a few months.

In the south of the continent, rain generally begins sometime in autumn, and most of it falls through winter and into spring. This is also the cold season, and reproductive strategies of wetland animals vary. The animals of ephemeral pools and seasonally flooded places, including some fishes and frogs, must breed as early as possible despite the prevailing low temperatures as the water will disappear within a few months (see Plate 3). Their young must grow and mature rapidly, or drought-resistant eggs must be produced to wait until the next winter's rains. Animals and plants in more reliably wet habitats develop eggs or prepare for flowering through the cooler months, and their reproduction is often triggered by rising temperatures.

In other southern areas, rain may follow a more or less predictable pattern over the year as a whole, but with occasional heavy falls at unpredictable times. This pattern is particularly common in coastal areas where storms may brew up in a matter of hours, and dissipate just as quickly. The rain from such storms sometimes penetrates a long way inland, and though it may be years between heavy falls in desert areas, this is often the only source of water for parts of the perennially dry inland.

The animals of ephemeral wetlands need regular dry periods, and many of their adaptations are directed towards survival when that inevitable dry spell comes. Some produce hard-shelled eggs, others form a cyst around themselves, also hard-shelled and water-retaining to preserve the minimal moisture they will get by on until the next rains. These dormant forms will survive far longer than just from one wet season to the next, often remaining viable even after many years or perhaps sometimes decades of drought. The great advantage of this strategy is that when they return to life it will be in a wetland with no predators, and abundant time to reproduce before predators arrive in any significant numbers.

All these more regular patterns are changing as global warming starts to bite. Rains in the north may become heavier yet more erratic, and in longer wet seasons their waters will spill south increasingly often. In the cooler south, the once regular south-westerly winds are less likely to bring rain, which is now as likely to come from the north though it is also becoming less reliable. Whatever the changes, extremes of drought, flood and even fire in some types of wetland are likely to become increasingly frequent.

Drought, flood and fire

Drought contracts the range of some wetland species, while creating new habitats for others that are pre-adapted to leaner, drier times. For those which are crowded into smaller spaces, drought is a testing time and selects strongly for those best suited to less than optimal conditions, while others with abundant new grounds available proliferate wildly while the chance is there. It is only as their new but temporary homes disappear again that the testing of the new genetic variations that appeared during the population boom will begin.

Selection during lean times may sometimes simply be for the most aggressive individuals or pairs, as in black swans. Over recent droughts I have watched unnaturally high concentrations of these far-ranging birds gather in the few local water bodies left that held enough plants as food for breeders, and later their young (see Plate 28). With time, the constant fighting among the overcrowded birds has driven enough birds away to match the available food supply to the number of breeding pairs remaining.

A three hectare wetland with dense water ribbon growth held the most crowded flock of 120 birds, with around 30 nests under construction at one stage. The wetland would obviously have been far too crowded if all of the nest-builders managed to hatch several young per pair, but over the next three months constant bickering and some serious fighting drove many birds off until only six pairs ended up actually sitting on eggs. Although not all of the other swans left, the others which remained had apparently given up on breeding for that season, or nested late in other wetlands when these finally filled after the autumn rains.

Most plants have little option but to remain put when drought sets in, and where this occurs frequently much of their biology is fine-tuned to this inevitability. Water ribbons (see Plate 28) are a particularly good example; tuberous-rooted plants which vary their growth patterns, quite literally, according to what the weather is doing from one day to the next. This might seem to restrict how and where they can grow, yet as a group they are among the most characteristic, well adapted and widespread wetland plants in Australia.

In times of drought all of their green growth dies down, leaving a fleshy crown and a fibrous root mat with numerous tubers buried underground to varying depths, different for each species or form. No green growth is produced until enough rain has fallen to wet the ground to some depth, and the relatively small new leaves often lie flush upon the ground, replenishing the underground reserves or perhaps even adding to them while the rains last.

It is only when the ground is actually flooded that the plant will draw seriously on its reserves, putting out larger leaves, and only if the flooding becomes quite deep that flowerheads and ultimately seeds will be formed. The seed doesn't have a long viability, relying instead on quick germination and formation of whatever tubers it has had time to produce before the waters disappear. Although seedlings don't have the bulk and drought tolerance of the mature plants, at least a few of them will survive from each successful seeding, to grow again if rains return within a year or two.

This drought strategy is so successful that some species have carried it further, adapting to even more extreme environments where few other plants are seen, apart from the odd annual blown in as a seed. Floodplain water ribbon is by far the most widespread species, found from New Guinea to north-western Victoria, in places which are dry for most of the year such as seasonal streams in the Northern Territory, or floodplains around the Murray River which remain dry except in the wettest years.

Southern forms of floodplain water ribbon must have survived extreme conditions in the arid inland of the last Ice Age, as they aren't found anywhere near today's coastal areas, and this suggests that they have always been an inland specialist. At Lalbert Creek near Swan Hill I dug out a dormant plant from an area which I had seen shallowly flooded six years before, but had been nothing but a dustbowl in the intervening years, to find it producing leaves within three days of being casually dropped into a bucket of water for the journey home. Northern forms of this species are visibly different from those in the south, and have developed their own regional responses to the short but intense wet season, including the ability to grow vigorously in as much as two metres of water, unlike the southern forms which prefer the shallows.

In south-eastern Australia, the pygmy water ribbon grows only in very short-lived pools, flowering and setting seed months before its larger relatives in deeper pools nearby, while narrowleaf water ribbon in south-western Australia thrives in even more ephemeral waters (see Plate 3), even on clay that bakes hard in summer. It often shares these pools with another ephemeral pool specialist, the south-western laceplant which also can grow, flower and even sometimes set seed in just a few weeks. (Water ribbons are also discussed as a major food plant for many wetland birds in Chapter 14.)

Flood is as much a part of normal wetland cycles in parts of Australia as drought, and this can also be seen in the behaviours and responses of inland animals. The occasional flooding of the great salt lakes that remain dry most of the time is the most familiar example, when they fill with exceptionally heavy overflows from the northern wet season. Whether they are saline or fresh, inland wetlands teem with life when newly flooded, particularly invertebrates and fish fingerlings whose parents were triggered into breeding mode by floodwaters, sometimes within a matter of days.

Rainfall also triggers new growth and breeding cycles in inland wetlands, not just in the terrestrial wildflowers which create a short-lived but spectacular show. Desert frogs emerge from burrows where they have remained immobile for years, to breed in the short-lived pools, and their tadpoles must mature into small but physically adult frogs in a matter of weeks. Inland strains of golden perch are different enough from those in the Murray–Darling that they may represent a different species – one that can mature eggs and milt and then spawn within several days of flooding. Though most of their offspring are likely to be trapped in drying lakes and other impermanent waters, enough will survive in new homes after the dispersal to keep the species abundant.

Even in less arid regions, some native fishes still show behaviours that are clearly adapted to distributing themselves widely during floods, at any cost. Driven by instinct, spangled perch (see Plate 1) swim manically upstream through shallow floodwaters even if they have to lie on their sides to do so, and are probably the origin of many stories about falls of fish with rain. Though they have no way of knowing whether they will find any water deep enough for survival when the flood recedes, and many individuals must perish as a result, the species remains widespread and abundant as a result of this strategy so it must be considered a success overall.

Fire doesn't affect most types of wetlands except during the most extreme droughts, but in recent years a combination of lightning strikes and dry conditions in the high country between Canberra and alpine eastern Victoria have set off some horrendous fires. Although these were a natural phenomenon, they were so severe that much of the limited sphagnum bog habitat of the corroboree frog has been destroyed. There has been some regeneration since, but this may have been the last straw for hopes of recovery of this already endangered species in the wild.

On the more positive side, it is reasonably regular fires that keep the extensive buttongrass plains of southern Tasmania open (see Plate 2), destroying competing vegetation (particularly teatree or sometimes paperbarks) which would otherwise shade it out and form the more familiar types of heathlands dominant on the mainland. Buttongrass is well adapted to these conditions, and flowers vigorously for a year or two after a cleansing burn has gone through. Yet in dry conditions, and if large amounts of peat have built up in heathland soil, fire becomes a purely destructive force, burning for years underground in drought conditions, and leaving nothing behind but ash.

Changing salinity

A generally low and worn-down landscape, sprawling over great distances, creates sluggish rivers and shallow lakes, often with a high salinity as the effects of hot summer sun and low humidity take their toll. Where lakes are the lowest drainage point in a landscape or catchment, they are inevitably saline, and sometimes

The birth and 'death' of Lake Colac

When I was a zoology student, the western district lakes of Victoria were still a paradise for budding limnologists, and I have studied them intermittently for nearly four decades now. Colac is our nearest large town and I have spent many hundreds of hours around its lake, taking photos and watching shifts in the biological players through a time of great and often dramatic changes, in a microcosm of the effects of drought across southern Australia (see Plate 13).

Left stranded in a large but shallow depression when the lava wall of the original and vast Lake Corangamite was breached, Lake Colac has come and gone over the centuries, as a lunette of wind-heaped sediments to the east testifies. With an average depth of not much more than two metres, and evaporation generally higher than the rainfall which fills it in good years, the lake would usually disappear after a couple of years of drought if it didn't take up to three years for groundwater from the north-west to drain into it.

In wet years gone by, the lake would overflow across marshy ground also to the north-west, or reverse the flow of its main feeder stream to discharge into the Barwon River, and this flushing effect helped keep salinity levels moderately low. Early studies of the lake's biota suggest it has changed little over the past century and more, with only a few larger sedges, reeds and at times water ribbons around the fringes, and small-fruited water-mat forming submersed thickets during times of high water. Although a few species of algae are sometimes abundant, particularly *Cladophora* attaching to everything projecting above mud level, free-floating phytoplankton aren't generally important.

The shallow waters were constantly stirred and turned over by wind over the approximately 25 km^2 of surface, so they were usually oxygen-rich, except in pockets where pollutants from industry and a sewage treatment plant were concentrated to create unusually high biochemical oxygen demands (BOD), as these were broken down by micro-organisms.

Since the beginning of the 21st century, the area has been in a state of more or less perpetual drought, and as the lake waters fell over the last decade, dissolved salts became increasingly concentrated. Eventually the increasing salinity killed off the introduced carp which were the dominant fish, and these washed ashore in their thousands (see Plate 13). By mid-summer in 2009 the lake had almost disappeared, leaving only a small, shallow and stagnant lagoon replenished by a trickle from Barongarook Creek.

As the waters first began to recede and the lake bed was increasingly exposed, seeds of marine clubrush and sharp clubrush germinated in huge numbers not far out from the old shoreline, and along with variable groundsel these covered newly exposed areas which had been too deep for plants before. It also presented an opportunity for expansion for other species of colonising plants which would usually have been hemmed in among established plants around the fringes.

Annual species with windborne seed appeared in countless millions on newly exposed sediments as the level fell further, with blown grass (known locally as fairy grass) producing such quantities of seed heads that it became a fire risk when heaped by the wind. Two species of willowherbs also proliferated, as well as the introduced aster-weed. The same patterns and often species were to be seen around most of the drying lakes in south-western Victoria. For all of these it was an opportunity to experiment on a grand genetic scale, with natural selection to choose the best new variants when their present home is flooded again one day.

With the return of rains in the following autumn the lake has partly refilled, drawing huge flocks of waterbirds at times, especially vegetarian species that feed on the abundant herbfields covering the mud. Though numbers of noxious fishes (particularly plague minnow and carp) have fallen dramatically over the course of the drought and have yet to recover, some native species have demonstrated their greater adaptation to the extremes of Australian climates. Landlocked survivors of spotted minnows and common jollytail (see Plate 13) spawned abundantly as Barongarook Creek began to flow again, and though both of these species usually only breed in estuaries, schools of thousands can now be seen in the creek, and they are probably the main prey of the pelicans that flock together where the creek enters the lake.

If global warming is indeed the culprit behind the current and seemingly never-ending drought, as much of the evidence suggests, then Lake Colac will only reappear sporadically during occasional wet periods over the next few thousand years, though it will become an ephemeral lagoon again each wet winter. Eventually, if regular rainfalls return it will be recolonised, not necessarily by exactly the same range of plant and animal species, though most of them are likely to be familiar. The question is whether there will still be any humans around to appreciate the return of the waters.

extremely so (see Plate 6). Although very few animals (most of them microscopic) tolerate such conditions, in the absence of predators or a wide range of competitors they are usually present in great numbers.

Insects decrease in variety as salinity rises, although flies and mosquitoes may take advantage of the reduced range of predators in more saline waters to breed up in impressive numbers. Some crustaceans will also multiply in numbers, particularly salt-loving copepods and ostracods. Freshwater plants flower less often and are less likely to set seed as they struggle to survive the salt tide, though they may make a temporary recovery in wetter years when the salt levels are diluted for a while.

Most inland waters are less saline, though there is often enough salt to restrict the range of animals present, and it will still have some effect on the flowering and seed set of freshwater plants if only by reducing their vigour to varying degrees. As most Australian freshwater fishes originally evolved from marine ancestors, salinity is usually not such a problem for them. Frogs are less able to adapt, and in increasingly arid environments will be under pressure not only from warmer and drier air, but also from fire events that destroy fallen timber that provides shelter and insulation against solar heat, and the ever-decreasing number of freshwater wetlands as evaporation concentrates salt in shrinking bodies of water.

Conditions like these must have come and gone over the past few millennia, perhaps many times. It could be argued that today's biota is made up of the highly selected, tough species that are most likely to survive further salinisation of inland waters, except that it isn't true. The vast majority of freshwater animals and plants simply died out in more saline areas in past aeons, and it was only their ability to spread to other wetlands which saved them.

Movement and change

Wetlands are a fickle habitat over long periods of time, evaporating in increasingly dry conditions, becoming ever more saline when they are the lowest drainage points in the landscape, or silting up gradually over centuries or thousands of years. Rivers change courses, cutting new beds during floods, and heading off in new directions altogether if the tilt of the land shifts even slightly during natural settling processes. A lagoon behind sand dunes may be fresh one year, but turn salt in the next as wind or waves create or close a passage to the sea.

Most wetland animals can obviously move on to some degree if their wetland home becomes untenable, and there is no mystery as to how birds and flying insects make their way to new homes. This is why so many native waterbirds are migratory, shifting over great distances routinely as a response to the unpredictable nature of our climates. Reptiles and amphibians are less mobile but can still travel considerable distances, particularly if there is an abundant selection of intermediary wetlands to move through, one step at a time, to a better home.

Most fishes have little choice about relocating as they have to follow the water wherever it goes, and adaptability is often their sole defense against change. Many Australian species respond rapidly to flood events, which open a window of opportunity to new habitats. Adults move upstream rapidly or spread into flooded areas around streams, breeding sometimes within a matter of days. Their offspring may be carried great distances to new potential homes joined briefly by floodwaters, though for most these become literally a dead end if they dry out completely.

Plants move more slowly than the proverbial snail, but the seed of many wetland species is equipped with downy parachutes or wisps that catch the slightest breeze. Though the seed itself may be tiny, it will grow rapidly if it lands on open, wet ground cleared of competing plants by drought. Others produce tough-skinned seed with a nutritious kernel, and though many of these are broken up in passage through waterbird guts, some come through intact to be deposited in a new home in a ball of the richest fertiliser known – a bird dropping. Still others have barbed seeds or a tangling filament, so they can hitch a ride in a bird feather or attached to its leg.

All wetland plants and animals must be able to abandon disappearing wetlands, and find their way to a new home by any mechanism possible, or they would not be here today. They also need to be remarkably adaptable when they do arrive, because in some cases it may a continent away, and in a different climate. The one thing they must all find at the other end of a successful journey is water, but the distances they need to cross to get there have become increasingly greater since 1788.

Part B

Invaders

4

Human impacts

The adaptability of native wetland animals and plants over thousands or even millions of years was emphasised in the last chapter, but over the last two centuries that adaptability has been put to new tests as indigenous plants and animals have been threatened by an array of processes created by us, the European invaders of a country we still seem to have little understanding or appreciation of.

These threats and challenges cover a wide array of ills from depletion of natural water resources to increasing salinity, with the effects of global warming only really beginning to bite now. Introduction of vermin and weeds also continues despite the environmental problems they cause, but these two ills are discussed in the next two chapters.

Drainage

The most frontal attack on wetlands is drainage, whether this is just a channel to run water off a low-lying corner of a paddock, or a massive system of canals such as have destroyed 90% of the wetlands in south-eastern South Australia, and possibly a similar area on the Swan coastal plain around Perth. At least one-third of all Victorian wetlands are gone, three-quarters of those in the Sydney area, and we don't really know how much has vanished in many other areas because no-one cared enough to note their passing.

The results have long been obvious. Storm waters run off to sea more rapidly, and carry more soil and nutrients than ever before (see Plate 17). Floods are more serious, because most watercourses have been altered so much that flows are less controlled by the natural processes which used to regulate them, including their surrounding wetlands which acted as a buffer that delayed the passage of flood waters and released them much more slowly than a concrete channel.

Wetland drainage continues to endanger still more of our fauna and flora, silts up rivers and underground waters faster, and has made Australia a lot drier, greyer and quite a bit less beautiful. In the more arid regions, wetlands must always have been scattered so that some animals only had the chance to move between them in exceptionally wet years, but since European settlement, many of these are now even more isolated, and the distances between all wetlands which have been significantly drained have become still harder to bridge.

While wetland plants and animals must always have been able to cope with a certain degree of change, in some places it is now probably impossible for many of those which can't fly, or hitch a ride with flying animals or the wind, to shift to new homes without human intervention. On another level, the reduced ability of some species to move between wetlands (due to greater distances between the remaining wetlands after drainage) also restricts genetic flow, and is likely to lead to more inbreeding in isolated populations, as well as effectively reduce the size of breeding populations.

Like just about any other source of potable water in southern Australia, groundwaters are also under increased pressure from water supply authorities. In many cases the water being used ('extracted' is the politically correct term) fell as rain years ago, and in many cases its origins may date to long before European settlement. Although the impacts of over-use of this resource are most evident on terrestrial ecosystems, drainage of surface wetlands has almost certainly reduced or put an end to recharge of many of these hidden waters.

Weirs, dams and other river 'improvements'

Weirs, dams and similar barriers are primarily used to bank up water for irrigation, livestock, and on major waterways to allow navigation by larger watercraft. Even a concrete 'apron' to allow traffic to cross floodwaters in drier inland areas can act as a shallow weir during peak water flows, and although it does not necessarily stop all fishes from migrating upstream when the urge takes them, it will make them much more vulnerable to predatory birds.

Weirs also have a number of incidental effects on the lifecycles of aquatic animals and plants, even on those which may originally been quite a distance from the water. Trees and shrubs such as river redgums and paperbarks grow in places which flood intermittently, and will weaken and eventually die if their soils remain wet for too long. Paradoxically, as water becomes increasingly scarce, many of the Murray River's redgum forests which were once in potential danger of drowning are now dying from lack of adequate seasonal flooding every few years.

Overcoming the effect of weirs and dams as barriers to fish movement is discussed in Chapter 7, but they also have impacts on the temperature of many rivers and associated waters, in turn affecting the ability of some aquatic animals to breed. The deeper waters behind a reservoir remain relatively cold throughout the year, often many degrees colder than the river they impound would naturally be over the warmer months. This is not necessarily a problem for animals living upstream of the barrier, as there are usually at least some backwaters and shallow inlets which warm up rapidly, and as these develop mature stands of vegetation may even recreate suitable breeding areas for some species.

Until recent times, however, overflow water from reservoirs has often been taken from the deeper waters, so temperatures for many kilometres downstream plunge. Even in the tropics, the lower reaches of a deep dam may be in the order of 10 degrees cooler than surface waters. This is not always a liability – for example, spiny crayfishes (see Plate 11) generally prefer cooler conditions, so in inland waters dwindling populations of the Murray crayfish may remain most abundant in the larger and deeper water storages.

Drawing discharge waters from closer to the surface solves the downstream temperature problem, but as with any problem caused by weirs this isn't always positive in its effects. The warmer waters are where most schooling fish and other species concentrate, so when the flow rates are greatest many of these may be swept downstream. Stunned by the fall over a weir, they attract predatory waterbirds such as pelicans and cormorants looking for an easy meal.

Irrigation dams compound the problems caused by impoundments, as they are usually drawn down during the warmer months which are the main breeding season for many animals. For example, freshwater catfish in the Murray–Darling build a nest of rounded gravel to raise their eggs above the silt bed, and will abandon the nest if the water level drops to any great extent. On the other hand, warm and shallow backwaters kept at a fairly uniform depth by a weir may provide an ideal breeding area for this species if fine pebbles are available.

Channelisation and the removal of snags from rivers have also had negative effects on a wide range of fishes, not just those that need free access to the sea or upstream. Snags rather than plants were once the most important form of shelter from cormorants and other predatory birds in many reaches of a river, apart from in backwaters where dense thickets of species such as eelgrass and pondweeds protect smaller species and fry from larger fishes. Some early descriptions of rivers like the Yarra suggest that it would have been difficult to get even the smallest boat any great distance past the fallen trees in the water.

Other effects of weirs and dams on riverine animals are less obvious. For example, the Darling River before European settlement acted as an artery allowing more tropical species such as purple-spotted gudgeon (see Plate 12) and the Murray rainbowfish (see Plate 32) to adapt to colder climates as they moved south over the past few thousand years. With the generally lower temperatures resulting from extensive damming along both rivers, the gudgeon has now disappeared from its southern range, while the rainbowfish has been broken up into increasingly isolated populations. It seems likely that this interesting experiment in adaptation has now been brought to an untimely end.

Irrigation

Until water for wetlands is regarded as a basic right (rather than being mostly directed to agriculture), billabongs and floodplain swamps are only likely to survive in a reasonably natural state with enlightened management, and then only if seasonal flooding can be arranged at least every few years. Yet water for maintaining wetlands is not given much of a priority, and one particularly large cotton farm pumps more water from the Darling River than is allocated to all environmental flows throughout the entire Murray–Darling catchment – only around 1% of the river's entire flow!

The greater part of the water running through irrigation channels is wasted, evaporating or seeping into the subsoil as it flows through open, earthen channels across some of the hottest and driest parts of this continent. Pipes are perceived as uneconomical despite these losses, and it is only recently that authorities have started to recognise that water is now an increasingly scarce commodity, and one of our most valuable resources, so its long-term conservation at all levels needs to become a serious priority.

Drought has caused some changes to the way water is used by agriculture, for example, restrictions on the growing of rice in many parts of the Murray–Darling catchment, and an increasing emphasis on dryland farming. It has been suggested that with increasing rainfall in the north, this is where we should be growing rice in the future, and though the original attempts to farm rice near Darwin were thwarted by magpie geese, there seems to be more prospect of growing upland (non-irrigated) crops if these receive adequate rainfall.

Rice is one of the most important food crops in a world where our stockpiled food reserves have been diminishing over the past few years, but growing other water-greedy crops such as cotton in hot, dry places seem a lot less justifiable. Even beyond the issue of the huge amounts of water this is drawing from already depleted rivers, it is also an environmentally dirty industry notorious for use of an impressive array of toxic chemicals – many of which will ultimately leach back into rivers and wetlands.

Clearing, grazing and toxins

Like most other types of indigenous vegetation, wetland and riparian (streamside) plants have been cleared on a dramatic scale since European settlement, and the resulting damage has been compounded by grazing the damaged areas. Poorly thought out clearing is a disaster no matter where it is carried out, but it is hard to believe that anyone could have thought this was a safe practice in places where floodwaters run at times.

The resulting erosion still goes on, with deep gullies now a prominent feature of many Australian landscapes, where they were almost certainly rare before. Repair work in these places where soils are well-drained or even dry for most of the year, yet scoured by floods cramped into a narrow channel at others, has become a major challenge for landcare groups and requires fast-growing trees and shrubs that will establish on very little water, yet endure being partly submerged at times.

Long before the notions of either not clearing along river and stream fringes, or of replanting them as quickly as possible with the species recently removed became obvious, a botched compromise became popular. The introduction of willows is now a separate environmental problem in itself, and the problems these cause and contemporary approaches to their control or management are discussed in the next chapter.

Grazing on wetland fringes and along rivers has theoretically been reduced because of the damage it does to re-establishing vegetation, and fencing along flood-prone areas is increasingly common. However, there are still many grazing leases in wetlands that need urgent protection, and these are often excluded from the protection offered by more recent legislation until the lease periods expire. In some cases, certain individuals even continue to run stock in the wrong places for short periods of time, for as long as they can continue to get away with it.

Clearing has also created new types of problems in estuaries and around coastal wetlands, particularly (but not exclusively) among mangroves, where iron sulfide soils formed in flooded soils are exposed to air by clearing, and oxidise to produce sulfuric acid (see Plate 17). Apart from the habitat reduction effect of plant removal, ground and surface waters become more acidic in these areas, and re-establishment of vegetation may not be possible in our lifetimes because the exposed soil becomes so acid that nothing will grow in it.

Particularly depressing are abandoned mud crab farms in northern Queensland, because these were approved and excavated at a time when comparable prawn farms in parts of Asia had already been abandoned due to acid sulfate soils, and the problem was already well understood. Mangrove clearing continues in places, fuelled by developers anxious to make their fortunes from people who love the sea, and wish to live near it. It is a tragedy that by paying for the conversion of mangrove and estuary into housing estates, they are helping to destroy much of what they would like to preserve – clean water, good fishing and the beauty of the coast.

The most subtle threat to aquatic life is an ever-increasing diversity of agricultural toxins and waste products from various other industries finding their way into rivers and wetlands. As these are not visible to the naked eye, they are usually only detected either by precautionary testing of the water (expensive and therefore avoided wherever possible), or when obvious physical, ecological or behavioural problems become apparent.

A recent example in the form of an unknown chemical in river water, causing a 90% hatch rate of two-headed bass in a hatchery on the Sunshine Coast in Queensland, triggered denials of responsibility by everyone who could possibly have been blamed in any way. Initially attributed to a macadamia farm that had been under suspicion years before, in the end no blame could be assigned to any one person or cause. Most such problems are less dramatic so they are simply not noticed, so if only 5% of the hatching bass had been two-headed, their deaths within days of hatching would just have been put down to a slightly higher than expected casualty rate.

Nutrients and dissolved oxygen

Wetlands are the lowest-lying areas of the landscape, and inevitably as water moves through them it carries all the soluble (and some less soluble) materials produced upstream. These include toxins sprayed onto farmlands, excess fertiliser, grey water runoff, nutrients and waste products from farm animals, and in the past also included many more serious pollutants. Even within the last few decades, many wetlands were still being used as dumping grounds and tips for just about any unwanted material, and the remnants of these can still be found around the fringes of some lakes and streams.

Excessive nutrient levels, particularly phosphates, potassium and nitrogen trigger the growth of algal blooms, as well as of other organisms that can deplete dissolved oxygen levels. Oxygen levels drop naturally as water warms up, but can also be dramatically reduced by surges of biological activity in nutrient-rich areas when these are suddenly flooded, triggering fish kills. Decreased flow rates and increased temperatures have also contributed to the proliferation of the often-toxic blue-green algae (technically cyanobacteria rather than true plants) over the past decade. Though these would always have become more common in drought years, they were usually present at relatively low densities so they often went unnoticed.

If environmental flows were increased so that natural wetlands refilled more often though less dramatically, these would partly help to alleviate the problems associated with nutrient build-up; it is not for nothing that created wetlands are built to improve water quality. Many aquatic plants readily take up some nutrients (especially those which trigger algal blooms) in quantities much greater than they need for immediate growth, and stockpile them against future shortfalls. These processes are discussed in more detail in the companion volume *Planting Wetlands and Dams* (2009).

Salinity

We have known for more than a century that large-scale clearing of native vegetation and irrigation are associated with ever-increasing salt burdens on the affected lands and water, yet the loss of large areas of salt-affected farming land (as well low-lying wetlands) has accelerated over the past generation. This problem is only going to become worse with global warming, as it becomes harder to re-establish suitable vegetation to restore some of the balance, on an increasingly meager ration of fresh water.

When slow-moving waters in hot, dry climates are slowed still further, evaporation rates increase and so the salt concentration must also increase. Many weirs are designed to bank up fresher waters so they can be diverted into irrigation systems, and as it flows through an extensive network of open canals or into dams, the evaporation rate increases still further. This is particularly obvious in some regions of the Murray–Darling where evaporation rates are estimated to be up to 80% in summer, and it doesn't take a mathematical genius to recognise that this means a fivefold increase in the concentration of dissolved salts.

While this may not seem dangerously high if the water was fairly fresh to start with, it is then used on dry land where the salt remains, accumulating generation after generation, until the soil is unfit to grow any useful crops. In history, this effect has caused the demise of whole societies dependent on irrigation in dry climates, and will put a 'natural' end to inland farming through much of the Murray–Darling in the long term. Even more objectionable is the use of irrigation waters to maintain green lawns and lush gardens in arid, inland areas that pride themselves on how many sunny days they have – an offense to both common sense and the environment.

Fishing, aquaculture and hunting

Although inland fisheries in Australia have always been small compared to those offshore, until a few decades ago they were making significant inroads into populations of popular food fishes, over and above the effects of river impoundments. Most commercial fishing has come to a halt now, and there is an increasing and laudable trend towards catch-and-release among recreational fishers, but spawning habitat remains threatened or poor quality in many inland rivers, and fishing pressure continues to have an indirect effect on wild populations of various species through restocking programs.

Wild trout have long been well established in every river system where they can maintain their own numbers, and most of the misguided hatcheries breeding and releasing trout to maintain fishing stocks have eventually had to concede that their efforts were largely futile, and have given this up. A more recent threat is from release of indigenous fishes for the same purpose, usually bred from a very small number of adult breeders whose genes then become unnaturally dominant among wild populations.

The problem is not just limited gene pools, but that in many cases these fish are descended from stocks that have been selected for aquaculture purposes, and after a few generations, like aquaculture fishes worldwide these become visibly different from wild populations. Their behavior is also often different, because there is little selection in hatcheries for fishes that avoid or even necessarily recognise predators.

I am not against aquaculture, and in an earlier book (*Sustainable Freshwater Aquaculture*, 2007) have already expressed my opinion that the freshwater sector of this fast-growing industry is the closest to already being ecologically sound in many ways, and would require the fewest improvements to become truly sustainable in the long term. But stocks of aquaculture animals including freshwater crayfish are not wild stocks and should not be mixed with them, because the goals of aquaculture and wetland conservation are not the same.

Hunting also continues in many parts of Australia, despite reduced breeding habitat for many waterbirds as drought becomes more frequent. I have considerable sympathy for responsible hunters who can identify and target species under no real threat, and who stick to bag limits and other game laws. There are, however, still quite a few idiots with gun licenses around, and I have been woken in pitch darkness in the early morning in the Barmah Forest by shooters illegally blasting away in the dark long before they could identify anything. This offense was added to later that day, when I was also shot at by an obvious drunk while inadvertently swimming near his decoys.

Climate change or global warming

Remarkably, despite decades of worsening conditions in probably the majority of inland waters, the causative problems continue to be ignored, and water extraction for projects of dubious economic value (when their real costs are factored in) is still being increased in real terms. All of this is taking place in an era where global warming is starting to bite, further reducing natural flows in much of inland Australia, while increasing flooding problems in the north.

I am not going to enter into debate over whether these changes are caused by climate change (in which case they are natural), or whether they are the result of global warming (in which case we had better respond very fast). The precautionary principle has increasingly been applied in recent years to all kinds of projects with potential to cause irretrievable ecological harm, and applied on a global scale makes it clear that even if rocketing carbon dioxide levels are only part of the problem, they are at least a part we should be doing something about.

For this reason global warming is the appropriate term to use, instead of hiding behind climate change until the mounting evidence becomes impossible to ignore. Most alarming is the assumption that we are dealing with a linear problem, yet if temperatures reach high enough to release the vast reserves of frozen methane presently locked up in melting tundra and on the ocean floors, we will experience global warming on a scale that beggars belief. If similar events in the deep past are any indication, there may be a real chance of average temperatures rising in the order of 10°C, as methane is an even more effective greenhouse gas than carbon dioxide.

The effects on wetlands would only be a trivial part of the resulting problems, but already evaporation in many parts of Australia is increasing, with increasing salinity as a corollary, and wetlands are drying out faster with less time for an increasing number of species to complete their life cycle, and sea levels are rising. It will not be long before many low-lying freshwater wetlands will become saltmarshes, and although the biota of our wetlands has survived similar changes in the past few thousand years, there are now far fewer refuges inland for them to retreat to.

This in itself is enough reason to be undoing the effects of immoderate drainage, holding back and conserving fresh water wherever we can, not only for the immediate improvement of habitats and a buffer against the increasingly dusty, grey world we seem to be moving towards, but also to give groundwater more of a chance to recharge instead of continuing to deplete it without making any plans for when it abruptly runs out.

Most readers of this book will probably be in agreement with all these remarks, but if you own or manage land with wetlands, ponds and streams, rehabilitating these and undoing the effects of artificial drainage is probably as important a step towards conservation of a damaged world, as planting of trees is towards absorbing part of the surplus carbon dioxide that is (quite literally) fuelling the problem.

5

Weeds and habitat

What is a weed? The simplest definition is a plant out of place, but that can cover a multitude of sins. One well-known gardening magazine recently included rainbow nardoo on its eccentrically chosen list of the 10 most serious aquatic weeds in Australia, on the sole ground that it is declared noxious in Tasmania, after a single and minor infestation was recorded from a short section of coastal stream 50 kilometres south of Hobart. Yet if some fool had not deliberately introduced this plant here it would never have established, and there is little chance that it will spread to other streams in a part of Australia far south of its natural range.

Rainbow nardoo also happens to be a significant native habitat plant across much of its eastern Australian range, a fact the author of the article did not appear to know. Nor did he mention the several hundred introduced wetland plants that are spreading fast in other parts of this country, including many serious weeds, some of which were still being offered for sale by advertisers in the very same issue of that gardening magazine!

A mistake like this is just ignorance, but even botanists don't always agree on what doesn't belong on this continent, let alone what has been introduced from other parts of Australia. I have already discussed in some detail the sorry story of waterbutton in *Planting Wetlands and Dams* (2009), a plant declared introduced by a clique of botanists without any evidence, and overlooking three papers published in the 1980s showing that its pollen was present in lake cores up to 10 000 years old. The situation for several other wetland plants formerly regarded as native is less clear cut, but in the absence of any solid evidence of post-European settlement introduction it is premature to make any decisions about their status until adequate historical and genetic research has been done.

It would take a substantial book to even describe all of the wetland weeds which have already established themselves in Australia, and those that have been deliberately or accidentally let loose by a wide range of people from aquarists to agriculturalists. Rather than produce a theoretical list of the 50 worst weeds, I have selected an array of examples from different climates with a view to looking at ways of either dealing with them or working around them.

The loose categories below are not intended as absolute guidelines, but rather as an approach to dealing with well-entrenched infestations. Regardless of the species and the degree of threat it poses to natural wetlands, it would be folly to not attempt to eradicate it completely if it is only present in small numbers, or has only just appeared in a wetland. Some control methods are discussed in *Planting Wetlands and Dams*, but chemical control methods are rarely recommended in the accounts below because I have seen little evidence that these are anything but a short-term expedient in most wetland situations, and because they also invariably have side-effects on animals and habitats which can last for decades.

Weeds of disturbed places

Many introduced weeds are best adapted to disturbed situations, colonising open ground in the absence of competing plants, but not usually invading relatively undisturbed wetlands. Such weeds may be abundant in open irrigation channels and rice fields, obstructing the flow of water and competing for nutrients, but this is mainly an argument against having open channels. Irrigation canals also act to spread such weeds, giving them their best chance of reaching other wetlands; these problems and excessive evaporation would all be solved by using pipes instead.

Arrowheads (see Plate 17) are among the most distinctive of such weeds in inland regions, and are difficult to eradicate completely because many of them grow mostly submerged, and some also spread underground through masses of tubers. The annual barnyard grass is even more fast-spreading, producing copious crops of seed and congesting open drains, yet is rarely seen in nearby wetlands unless these are being grazed or otherwise regularly disturbed.

Starworts also self-seed readily, and may choke open drains if nutrient levels are high, but can usually be relied on to mostly die out as competing plants establish themselves. Clovers may establish themselves on moist to wet soils, but even in pastures must be replanted from time to time to keep them going, and although they are visually unattractive growing on the fringes of a wetland, they will generally disappear once taller plants shade them out.

Parrotfeather is similar to some indigenous water milfoils, but grows most densely in nutrient-rich urban streams. Once nutrient levels fall, it becomes much sparser and even large quantities can be removed manually. Fortunately, only a female clone of this species is present in Australia, so it spreads only through deliberate planting of cuttings.

Weeds ecologically equivalent to some natives

Some weeds are more or less ecologically equivalent to indigenous species, in terms of habitat for particular animals, and may not warrant the expense and effort of control if their populations have already stabilised. Drain sedge is a particularly good example which has spread at an alarming rate in eastern Australia, not just in wetlands, as it can even be found in a stunted form thriving in cracks in concrete!

This is a plant which warrants every effort being made to contain it if it is newly arrived, but if it has become abundant there is probably nothing that can be done. Each plant produces tens of thousands of seeds each year, and these will sprout rapidly if the ground they are in is disturbed by pulling out the adult tussocks. It does not stand out visually as it is not unlike some natives in appearance, particularly tall flatsedge, and as it is non-toxic and doesn't usually made inroads into undisturbed wetlands, the cost of controlling it in disturbed sites does not seem warranted.

Canadian pondweed and its relatives are closely related to the native water thyme, itself regarded as something of a weed in canals. All of these submerged plants are eaten by some waterbirds, and as they break up readily and every small piece is capable of growing, there is no prospect of ever getting rid of them once they have appeared in a wetland or (more usually) stream.

Alligator weed is also closely related to some tropical natives, and as the only clone present in Australia doesn't set seed, its spread to-date has largely been through Asian communities growing it as a food, though education about its weed potential has apparently solved this problem. Biological control with a beetle has been reasonably successful on plants growing in water, but is not as effective on the more stunted plants found on moist soil, and the main management issue is preventing its introduction to new areas.

Although there are a number of indigenous waterlilies primarily in the tropics, introduced species deliberately planted into ponds and dams are more often seen. The common blue species seen in Queensland farm dams is usually the cape waterlily, and as this spreads rapidly from seed it has become a minor weed in places. Control is difficult once enough plants are present and probably pointless if it is widespread in the surrounding area, and should not be attempted unless you have positively eliminated any possibility of the plants being any of the native species.

In southern Australia most naturalised waterlilies are hardy hybrids, many of which don't set seed so they usually don't spread far from where they have been deliberately planted. Although out of place in wetlands they cause little harm in farm dams and even offer the same benefits as indigenous species further north – shade, shelter and reduced water temperatures in hotter areas. The most worrying exception is the Mexican waterlily, still sold as an unnamed yellow ornamental variety at times, though it is far from free-flowering and produces rank masses of crowded leaves from tangled masses of rhizomes which can put up leaves from three metres underwater. This is one waterlily it is worth bringing in an excavator to deal with!

Water hawthorn is regarded as a minor environmental weed wherever it is found, but over 150 years in this country has rarely spread more than a few hundred metres from the places it has been planted, and mainly thrives in urban settings as it is eaten by a variety of waterbirds. Like the native laceplants, some of which it resembles, this is a drought-tolerant species that dies back to an edible tuber in hot weather. A number of populations seem to have disappeared in the face of continuing drought over the past decade, yet it may be premature to treat it as a relatively benign weed as some reports suggest it may be spreading more actively in places with increasing summer temperatures.

Coarse clubrush looks like a larger form of the native common clubrush, and is a common weed mainly in disturbed situations in the central New South Wales area, though it is capable of invading natural wetlands too. As new plantlets grow from the flowerheads, it is almost impossible to remove manually once it has formed large enough stands, but is not difficult to deal with if still localised.

Weeds we have to live with

Some weeds should ideally be controlled, yet are unlikely to ever be eradicated, the best example being watercress which is found in probably every cooler stream of southern Australia where it could possibly thrive. The thickets it forms in shallow waters seem to be acceptable habitat for at least some frogs and smaller fishes, and the peppery leaves are eaten by humans and waterbirds. Although it is easily pulled out, as with many other brassicas a mature clump will usually have produced hundreds of thousands of small, hard-skinned, long-lived seeds, and these will keep on germinating for at least a decade after being buried in mud.

Most rushes in Australia are indigenous where they are found, though it is likely that they are much more abundant these days as a result of grazing and removal of competing plants, and although they have some habitat value they are often an indicator of degraded conditions. Their fine seed is copiously produced and germinates freely on exposed clays, and the plants are unpalatable enough that even cattle will only tentatively browse on them when nothing much else is available.

While the native species are acceptable enough in a wetland, there are also a few introduced species that are equally difficult to deal with. Some of these are annuals and once established will keep on coming up for many years, while others are more perennial and are so widespread now that there is little prospect of eradicating them from hundreds of kilometres of river flats and previously grazed wetlands. These are probably ecologically similar to various native species they closely resemble, and are sometimes difficult to tell apart. The worst of these is sharp rush, a spiny-tipped relative of the native sea rush which readily colonises saline soils. A slasher mounted on a tractor will easily destroy tops of the adult plants, the shallow rhizomes remaining are easily lifted with a mattock, and seedlings are easily pulled up once they are large enough to grip.

Cat-tail is a dark-pokered version of the native cumbungis (see Plate 10) and spreads just as readily by windborne seed. Small outbreaks can be pulled out by hand, but once a large stand has formed there is little chance of managing anything more than the outside perimeter, and this species also seems to spread from seed much more readily than the native forms. Habitat values of cumbungi are discussed in Chapter 10, but in all other respects these are ecologically closely equivalent to, and not much less weedy than cat-tail.

Water couch has been regarded as a native for most of its tenure in Australia, but is more recently being treated as a weed. In a practical sense it doesn't much matter which it is, as the dense scrambling mats it forms in high nutrient situations are undesirable because they smother most other wetland plants from near the water's edge to the shallows. On the other hand, in more natural conditions this grass is rarely a vigorous grower, and can be ignored unless there is little enough of it that it is possible to remove by hand.

Floating weeds

When floating plants grow too thickly they reduce oxygen supply to the waters below, so only a limited range of particularly tolerant invertebrates are likely to survive (see Plate 31). Excessive growth is more usually a symptom of high nutrient situations, and the worst infestations are likely to be seen in urban and semi-urban areas, with the partial exception of water hyacinth – the world's worst aquatic weed – which can blanket even the largest rivers. This is still sometimes sold as an ornamental plant even though it is banned throughout Australia, and though primarily tropical it grows rampantly as far south as Victoria. Salvinia is almost as weedy and similarly adaptable, though attempts at biological control in the form of an introduced weevil have been more successful with this species than in the case of water hyacinth.

The most conspicuous native floating plants are fairy ferns, forming red to green mats which can smother the water surface in high nutrient situations, but these are usually only problematic in farm dams. They are present in virtually every wetland, usually in limited numbers because long-established populations are controlled by a variety of viruses and pests. This can be a key to their control elsewhere if necessary, and introduction of a few handfuls of unhealthy-looking plants of the same species will sometimes have dramatic results if the viruses transfer successfully.

Water lettuce is regarded as introduced in tropical eastern Australia, though possibly indigenous in the Northern Territory, and although it can be invasive in high nutrient conditions has been successfully controlled with a burrowing weevil.

No-compromise weeds

A number of weeds are seriously invasive species which must be controlled if any semblance of ecological integrity is to be maintained. If any of these are present, get onto their control as an urgent priority because these are problem species that will become worse very rapidly. Some of the most widespread of these were introduced as pasture grasses for wet places, where they have become rampant and smother every other living plant other than the tallest shrubs.

Para grass is a tropical to subtropical species that is completely out of control in parts of Queensland, spreading into wetlands from drainage ditches, and forming choking mats in shallower waters. Eradication is a high priority for this species from the earliest stages, because once established it will become a perennial problem that even herbicides can't control. In the Northern Territory, Hymenachne grows even deeper, and in its spread across many floodplains chokes out watercourses and even native waterlilies. Why reed sweetgrass and several of its relatives were introduced is not known, but in southern Australia these are as intractable a problem along slower-moving streams and in nutrient-rich backwaters as the pasture grasses, with the added bonus that they are toxic to stock for a part of spring.

Pampas grass was certainly introduced as an ornamental, but despite its common name grows best in wet soils and self-seeds readily on the less waterlogged soils of wetland fringes as well as on higher ground. Large tussocks are wiry and tough, almost impossible to dig out, but at least herbicides can be precisely applied within their crowns to kill individual plants, with little chance of spray drift. Burning will also kill individual plants, but may have to be done several times.

Peruvian water-primrose is mainly a problem weed around Sydney, though it is also becoming increasingly abundant further north. As it grows large enough to shade out most competing plants and then sheds its leaves in colder weather, the denuded wetlands it creates offer little in the way of habitat for any indigenous animals. Of most concern is that it could potentially spread across all of northern Australia, but control is difficult as its seed is reportedly spread by birds.

Willows are one of the most common riverine weeds of southern Australia, introduced to hold stream banks together after the native vegetation was cleared! They do a lousy job of it, being brittle and soft-rooted, with any snapped-off pieces taking root wherever they wash up downstream. Unlike native trees, they shed their leaves all at once in autumn, providing little in the way of food or shelter for aquatic animals. Willow control has been discussed in some detail in *Planting Wetlands and Dams*, and is another case where a precise application of herbicides can be justified in control programs. However, the root systems should not be grubbed out once the trees are dead, but should be allowed to decay gradually, while planting replacement riparian vegetation as the long-term protection against erosion.

In the south the mostly introduced blackberries, and in the north the even more vile lantana cover thousands of kilometres of stream bank, growing over everything but the tallest trees. Blackberries are also good shelter for vermin including rabbits and foxes, making natural recovery of indigenous plants unlikely, but can be shaded out gradually by planting with dense-growing terrestrial trees such as blackwoods. Lantana is more shade-tolerant, and generally requires a lot more initial clearing or slashing.

Arum lily is at its worst in Western Australia, where it lines some streams for many kilometres, with little else growing among it. The plant itself is toxic to many animals, as well as unpleasant smelling so it is not even attractive as shelter. Also widely planted in many parts of eastern Australia, this species has the potential to become an ever more prevalent weed in new places as global warming takes its toll.

Mimosa (or giant sensitive plant, not an apt name given its prickly nature) is a noxious weed of the tropics, covering considerable areas in the Northern Territory and spreading by seed carried in segments of its floating pods during the wet season. The spiny thickets along streams make the water almost unapproachable for larger animals, though water buffalo help its spread by opening trackways through other wetland vegetation so that the seed can germinate more easily with reduced competition.

Indigenous plants as potential weeds

Plants growing within their natural range are not weeds, but many Australian wetland species have real weed potential if grown elsewhere. Paperbarks are the prime example, and the broad-leaved paperbark of eastern Australia is regarded as one of the most serious and invasive introduced plants in Florida, USA. Ironically, at another extreme the salt paperbark is both regarded as vulnerable within its native range in Victoria, and also as invasive as it is extending its range, presumably through seed escaped from ornamental plantings!

Even among terrestrial plants, many Australian 'natives' planted outside their natural range are probably a greater threat to natural ecosystems than the majority of introduced plants. They are not only already well-adapted to conditions in this country, but also have real potential of establishing themselves on a grand scale, especially if wildfire clears the way for them. Many aquatics don't even need to have competition removed to entrench themselves, and for this reason any indigenous plant grown near natural wetlands where it does not occur naturally should be regarded as a potential environmental weed.

6

Alien animals

One of the earliest preoccupations of European settlers was trying to make an unfamiliar land more like the homes they had left behind, and apart from the many weedy plants they introduced they also deliberately brought in a range of undesirable animals. It is hard to imagine what they had in mind in many of these cases – rabbits were already a notorious pest in Europe, while foxes were efficient killers of any small animal from poultry to lambs. It is still harder to imagine anyone in contemporary Australia who can be so brain-damaged as to deliberately and illegally introduce foxes to Tasmania, or new cichlid fishes to subtropical rivers, yet these people are out there and they are still doing their worst.

Birds

With plants it isn't always clear as to what has been successfully introduced or has arrived under its own steam but the case is much clearer for animals, with one exception. Cattle egrets (see Plate 26) were introduced deliberately into northern Western Australia in 1933, recorded next in large numbers in Arnhem Land 15 years later, and first sighted in Victoria a year after that. They are now the most abundant egret across most of Australia, and although they form colonies with other egrets in wetlands when breeding, are just as likely to be seen wandering in large groups among cattle.

Whether the introduced birds contributed much to the rapid expansion of this species is a moot point, as it was spreading worldwide at an equally impressive rate throughout the 20th century. There is no reason to think cattle egrets could not have reached Australia across the small distances from Indonesia and south-east

Asia (they had reached New Zealand by 1963 without help), but there is considerable doubt as to whether they would have established without widescale clearing and farming of cattle.

Considered as a possibly introduced species, cattle egrets are relatively benign and make an attractive addition to flat paddocks studded with cows, which is fortunate as there is no prospect of getting rid of them. It is possible that they compete to some degree with other egrets for nesting areas among paperbarks and similar trees, but certainly not for food as the cattle egret prefers to feed in open lands, specialising in the insects stirred up by cattle and other large animals, or even hunting actively for grasshoppers.

The only other introduced waterbird of concern is the mallard, the males of which are aggressive and brightly coloured, and have the potential to cross-breed with native black ducks. Despite these concerns, mallards remain uncommon outside urban areas, and are no more aggressive than their domestic descendants which are often seen in their hundreds on some ornamental lakes. Mallard–black duck hybrids are even less common, and in 30 years I have still only seen these in captivity.

For nearly all other introduced animals there is no ambiguity – they have been deliberately brought in for aquaculture, ornament or for hunting, while in some cases feral populations such as wild pigs have established from domestic escapees. Few of the introductions discussed below are relatively innocuous unlike cattle egrets and some weeds which we can learn to live with, and many of them have altered wetlands on a grand scale through competition, destruction of plants, spreading of serious weeds, and in one case through simply poisoning most of its potential predators.

Introduced fishes

Some introduced fishes are rarely seen because they have shown little sign of spreading a century or more after their introduction, so tench, roach and brook char are not discussed here, while the two species of salmon don't breed in natural wetlands, and populations are maintained only by deliberate stocking of hatchery fishes into some southern lakes. Many other introduced fishes are more widely distributed, including an ever-increasing list of feral aquarium fishes which have been deliberately introduced in some cases.

In cooler areas, several species of **trout** have been deliberately stocked for fishing purposes, and two of these are now well established in the wild. Brown trout are particularly successful, moving from one coastal river to another via the sea. The more warmth-tolerant rainbow trout is also breeding in some areas, but the brook char remains very localised despite past attempts to establish it more widely. In Australia, these are primarily fishes of mountain streams and deeper lakes, as they don't tolerate high summer temperatures in shallow waters, and they only breed in cool, oxygen-rich, flowing waters with a sandy or gravelly substrate.

Trout are active predators and have undoubtedly reduced the numbers of most native minnow (galaxiid) species, and probably also barred frogs that breed in backwater pools in cooler areas, as well as affecting larger natives such as blackfish that were significant natural predators in cooler streams before European settlement. Some of the most striking and localised species of galaxiids are now potentially endangered as a result, and attempts have been made to protect them by constructing trout-proof in-stream barriers, after which all trout were removed from above the weir by electrofishing. Unfortunately, in some cases trout have since been reintroduced above the barrier by fishermen with a blank indifference to anything but their own hobby.

The most widespread and abundant exotic fish is the **plague minnow** or gambusia (see Plate 19), formerly known as mosquito fish as it was introduced to control these pests, though it has now become obvious that they are less effective at eating the wrigglers than many indigenous species. The family it belongs to is sometimes known as toothcarps, and they do attack the fins of other fishes and tadpoles. As they are present literally in plague proportions in many places, they are also serious competitors for food, and are probably part of the reason some indigenous species such as the purple-spotted gudgeon have largely disappeared from southern parts of their range.

There is little chance of getting rid of plague minnows once they are present, let alone controlling their numbers in any but the most limited way as it is a livebearer, large females giving birth to hundreds of live young over the warmer months. Keeping them out of any habitat intended for native fishes or frogs is essential as it only takes a single pregnant female to establish a new population, and she can produce several batches of young from a single mating. Although widespread through south-eastern Australia, their distribution is more restricted in Western Australia, and the few isolated populations in Tasmania have hopefully been eradicated or at least contained for the moment.

Other more tropical toothcarps including **guppies, platies** and **swordtails** are also established in many warmer areas, especially around urban areas where their prolific breeding overcrowds aquariums, so surplus fish have been repeatedly dumped into natural waters. **Goldfish** are another widespread aquarium fish, and are often abundant in river systems as a result of dumping, reverting from the popular gold and reddish coloured forms to the better-camouflaged bronze and green wild forms. Apart from being yet another competitor for native fishes, these are also carry exotic diseases which may have helped reduce populations of some native fishes.

Their larger relative, the **carp,** has had more dramatic effects on the ecology of slow-moving rivers and lakes over much of southern Australia, roiling bottom sediments so water becomes more turbid and some plants grow poorly or not at all, while eating or uprooting others. Both carp and goldfish are prolific breeders, and are better adapted to tolerating the disturbed conditions they create than many indigenous species.

Redfin perch were introduced as a game fish, but although they can grow to a considerable size in ideal conditions, in most waters they remain stunted and breed at a small size. Small and bony as these runts may be, they are aggressive carnivores that feed indiscriminately on a wide array of invertebrates, as well as any fishes or frogs that will fit into their large mouths. Their willingness to breed in farm dams led to their being widely stocked several decades ago, so they are now found through most southern river systems.

Several species of the primarily tropical cichlid family have become established in warmer parts of Australia, the most prolific of which is so-called **tilapia**, one species of a large group many of which may be referred to by this name. Breeding from an early age, it can establish itself rapidly and is probably a serious competitor for native fishes, and being partly herbivorous it can cause as much damage to many plants as carp. Other, more cold-tolerant species of cichlid are also appearing in small numbers in cooler areas as far south as Perth, but these are usually eradicated not long after being reported, and whether any of them have established permanent populations is unclear.

Part of the problem with trying to stop new fish introductions is the aquarium trade, or more precisely, people who buy aquarium fishes and later dispose of them in a nearby river or swamp. Australia stretches from cold-temperate zones to the tropics, a span of climates and places potentially offering suitable habitat for a wide range of introduced species. Fortunately, most people who discard unwanted fish don't bother to do any research as to where they are most likely to establish themselves, and the majority of aquarium fishes still allowed into the country have specific needs and are unlikely to thrive in most places they are likely to be dumped.

There are exceptions which have yet to be found in the wild, but are well adapted to a wide range of conditions. The Chinese or **half-banded barb** is a close relative of the all too successful carp tolerating temperatures down to 5°C. Half a dozen young fish tested in one of my larger ponds produced a swarming colony of around 600 breeding-size fish within 15 months. This was no problem in the nursery as the test ponds are designed to empty into dead-end sumps from which nothing can escape, but if this fish is introduced into any southern river or lakes there will be nothing to stop its spread.

Several other barbs are also locally established in warmer parts of Australia, and other species such as the **spotted livebearer** (a close relative of gambusia) have started to appear in new places. Although we can hope that this will do nothing that gambusia hasn't already achieved, there is already some evidence that it can outcompete even the minnow.

The **oriental weatherloach** is another aquarium fish which has spread rapidly in the past two decades in southern waters, and research on it is still at the stage of identifying just which of several possible species have established in natural wetlands. It doesn't attack other fishes and is generally an inconspicuous addition to our fish fauna as it hides in mud during daylight. In a set of trials I carried out with a variety of native fishes, however, weatherloaches ate almost every egg laid in the floating vegetation of a rainbowfish-breeding pond – hovering near the surface and devouring the eggs within seconds of them being laid.

Aquarium fishes released into natural waters may also introduce new diseases and new strains of diseases from overseas. Some of these may be implicated in the extinction of a number of frogs as discussed in Chapter 13, while others may be affecting some native fishes. Among those known to have been recently introduced is epizootic ulcerative syndrome, a fungal infection also started to spreading in New Guinea and parts of South-East Asia, but even such common and widespread fish diseases as white-spot may have been introduced through the aquarium trade before anyone had thought to keep records. This may explain the extreme sensitivity of several native fishes including spotted minnows (see Plate 13) and nightfish to this disease, and may be why they are mainly restricted to flowing waters where the short-lived, planktonic disease larvae can't build up in numbers because they are constantly being swept downstream.

Controlling feral fishes

Unlike the many indigenous fishes which have suffered considerable reductions in range, or are even threatened with extinction, feral fishes are mostly extremely adaptable species that have rapidly spread through as much of their potential range as they have been able to reach. In part, this has been through deliberate introductions, but live fingerlings and smaller fishes used as bait are also often released by anglers, who should know better than to introduce these potential competitors for their favoured fishing species into new waters.

There is little or no prospect of eradicating such vermin once they are established except in small water bodies that can be at least partly drained, and though their impacts on native species and frogs in particular can be serious, at least they make good food for some other animals. Most freshwater turtles don't care whether they are feeding on plague minnows or rainbowfishes, while herons and cormorants are probably as happy to make do with carp as with the tadpoles they often displace.

Management and control issues for feral fishes come down to two things – whether they can be kept out in the first place (or after being exterminated), and whether there is any point to controlling them once they are present. In turn, this brings us back to the central question of what the wetland is being managed for, so if maintaining a breeding population of long-necked turtles is a primary goal, swarms of plague minnows may actually be a bonus as a food fish, outbreeding many of the native fishes they have displaced.

The problems of keeping unwanted fishes out of some types of wetland have been discussed elsewhere, but if they can be screened out permanently it will be worth poisoning off any vermin species already present. Before using any kind of poison or a chemical that has comparable practical results, all native fishes should be removed, or at least as many of them as is possible. In smaller pools this can be done with nets of various types, but in larger areas electrofishing (requiring a permit and trained personnel) is more effective, and will also provide an opportunity to thin many of the vermin as well.

Poisoning has often been done with rotenone, which breaks down to harmless substances within about two weeks, and usually only affects fishes rather than other animals. It can only be used under permit and is becoming increasingly difficult to obtain, as well as expensive and not always effective in larger bodies of water because of the difficulty of mixing it in uniformly. A concentration of 0.5 parts per million (ppm) of rotenone will kill fishes in many cases, but this should be increased to 0.75 ppm or more in harder, alkaline water.

Other treatments that don't require permits in most cases include agricultural lime (*not* quicklime) applied at around 100 kilograms per 20 000 litres of water, to rapidly raise the pH to around 9, assuming an initial pH of around 6 to 7. A slow increase may allow vermin fishes a chance to adapt, and it may be better to aim closer to a pH of 10 to make sure of a complete kill. Swimming pool chlorine or sodium hypochlorite will also work at a concentration of 4 ppm, and will dissipate to harmless levels within a few weeks.

The main problem with applying any such treatments is mixing them in, as this must be done as quickly and uniformly as possible. They are easier to apply (and less of them is needed) if the dam or wetland already has a drawdown system built in so that the water level can be lowered considerably, or if it can be partly pumped or siphoned out. The smaller the resulting pool, the easier it is to remove any native fishes present, and although it may be tempting to try to net out all the vermin as well without a follow-up chemical treatment, keep in mind that a single male and female pair of survivors remaining as fry or fingerlings is all that is needed to re-establish them. A fire pump with the hoses used underwater to create strong currents is a simple aid to mixing, as is an outboard motor mounted on something solid in the shallows.

Amphibians and reptiles

Cane toads (see Plate 19) are among the very worst animal invaders of Australian wetlands, and have been spreading inexorably across northern Australia for many decades now. With time and global warming, they will also probably move a lot further south, some models for their climate tolerances suggesting perhaps as far south as Sydney. Originally introduced from central America to control beetles and other insects in sugarcane crops, they have proven to eat just about anything else by preference.

On the one hand, cane toads are large and can swallow surprisingly large prey including diverse native frogs, while on the other their skin produces toxins that kill naïve predators from quolls to goannas, as none of these have had any reason to suspect there is anything wrong with eating anything frog-like they come across in the past. As the toads are large and clumsy, there is a prospect of keeping them out of some wetlands with a dense fringe of vegetation, particularly if this includes sedges and other tussocks with sharp-edged foliage. Hunting them down and physical removal is unlikely to ever keep up with their considerable breeding capacity, while currently popular exercises such as treating them as golf balls are just a form of sadism.

We can hope that populations of at least some native predators will recover and rebuild from individuals that are either immune to the toxins, or aren't tempted to sample them, and there will probably be ways to reduce their impacts and even to discourage cane toads consistently. At the present time, however, most of these things are just ideas on the horizon, conjectures and often just idle speculations. It would be a positive thing if the ever-increasing number of people living in affected areas put more time into experimenting on ways to reduce the effect of this scourge, and less into complaining that scientists should be doing something about it instead!

Several species of **newts** have been kept in Australia for many years, until they were banned more than a decade ago as they also secrete skin toxins which may endanger predators. Although they are probably much more fussy in their habitat requirements than toads, the places and climates they are best suited to are cooler and wetter parts of southern Australia – the perfect complement to cane toads should they establish themselves.

Red-eared sliders are an introduced turtle originally from around the Gulf of Mexico, but now on the World Conservation Union's list of the top 100 invasive species. Although banned from import, an unknown number seem to have been smuggled in over the years, and have apparently been dumped into waterways once the bright red and yellow stripes of younger animals fade as they grow, so they become less attractive as pets. So far they remain relatively uncommon in eastern Australia, but as more animals turn up it is becoming clear that they are probably breeding in some places, and females can live for 40 years, laying up to 70 eggs annually. The adults are retiring and not often seen, and if observed should be reported to State fisheries authorities, as they require specialised trapping methods to locate and remove.

Foxes (and rabbits)

Of all the introduced carnivores, foxes are by far the most destructive of wetland wildlife. They are an adaptable predator which will readily dig up turtle eggs, kill nesting birds, and take any other smaller animal as opportunity arises. Although they can be controlled to some degree with poisoning programs using 1080, to which many marsupials have a much higher tolerance than introduced mammals, only fencing will keep them out of critical breeding areas.

Some fence designs are too elaborate and expensive for any but the best-funded wetland projects; for example, those used to protect endangered western swamp turtles near Perth. I have had good results from heavy-gauge rabbit mesh (foxes may chew through thinner wire if hungry enough) 1.0 to 1.5 metres high as a basis for a relatively inexpensive fox-proof fence, also stopping the rabbits it is specifically made for. To discourage digging, the mesh should be attached to the posts with an outward curve at the bottom at least 15 cm wide.

Except on the heaviest soils, an additional panel of rabbit mesh 30 cm wide (another standard width) will be needed, tied to the outside edge and pinned down with wire staples until grass grows through it. Foxes don't seem to recognise that they could still dig under this, if they move back from the fence before beginning to dig. A strip of bird mesh around 30 cm high attached at ground level will keep very young rabbits from pushing through and establishing breeding colonies on the inside. A rabbit colony on the inside will not only attract foxes in its own right, but as the rabbits will also be able to dig their way *outwards* with no difficulty, they will create obvious entries for the predator.

The fence should be extended upwards by attaching an overhanging mesh strip *at least* 30 cm wide along the top edge of the rabbit mesh; this can be a lighter gauge mesh as it only needs to stop foxes climbing. Tie-wire lengths are used to hold the overhang loosely away from the fence, as foxes' claws can't grip onto shaky mesh, but it is essential that no trees or shrubs grow too close to the overhang as foxes can climb these to get over.

Water buffalo and feral pigs

If there are water buffalo or feral pigs in or near your local wetlands, it is to be hoped that you like shooting powerful guns, because that is what management of these two pests is about. Water buffalo are mostly confined to the Northern Territory, descended from domestic animals gone wild, and their numbers would probably be managed more effectively if they were treated as vermin rather than as an icon of the territory. In other areas where they are seen, they are usually shot quite rapidly as a desirable trophy with no legal restrictions on its destruction.

Water buffalo are dependent on good water supplies, and while they spread out over greater areas during the wet season so the harm they do is more diffuse, as floodwaters recede they concentrate increasingly around remaining waterholes. Although they could be more effectively managed as semi-domesticated animals and their meat is a valuable export commodity, the damage large groups can cause in the relatively limited range they forage over changes the nature of vegetation in wetlands, and opens up channels and disturbed soils so these can be more readily colonised by serious weeds such as mimosa.

Feral pigs (see Plate 19) are much more widespread, much harder to hunt down, and cause a much wider range of problems than buffalo. Despite the cute reputation piggy-wigs have among city dwellers, among mammals, pigs are the single most serious agricultural pest, and boars and females with young can be dangerous to humans if cornered or disturbed. They also carry a number of serious diseases some of which are transmitted to humans via mosquitoes, while others make them unfit to eat, so hunting them for meat is not an economic proposition.

With their wide range through low-lying areas and wetlands as far south as northern Victoria, pigs are also potential vectors for many other serious diseases which could enter northern Australia from Asia. These include rabies, foot-and-mouth, and potentially new forms of swine fever presently confined to pigs alone, but which could become the source of as yet unknown diseases.

Pigs also destroy many edible wetland plants, which is not a problem for the more resilient species, but they have the potential to eat out what is left of native populations of increasingly rare plants such as the giant fern, and along with water buffalo are implicated in changing the structure of river pandanus thickets, a significant habitat in the Northern Territory. Whatever your feelings about the shooting of animals, feral populations of this animal have nothing to recommend them, and anyone who is willing to hunt them through swamps of disease-carrying mosquitoes should be encouraged.

Alien natives

As has already been emphasised in the previous chapter, the terms native and indigenous are not synonymous, and an Australian animal introduced outside its natural range is just as much an alien as anything from overseas. As these may also be to some degree pre-adapted to Australian conditions, they may potentially be more of a threat to true indigenous populations than most non-Australian equivalents. Some native animals such as the galah may have expanded their natural range as a result of changes brought about in the past two centuries, but only deliberate and accidental introductions are considered here.

These include marron from Western Australia and the common yabby from the mainland onto Kangaroo Island in South Australia, both of which have escaped from aquaculture ponds into virtually all natural waters. Yabbies are by far the more abundant of the two species, and in combination these are now by far the largest and most abundant invertebrates on the island, feeding on a variety of foods from vegetation to any smaller animal they can capture. While there has been little in the way of studies of aquatic environments on the island before their introduction, there is no doubt that the aquatic ecology of the island has been changed irretrievably.

Striped grunters are among the most recent additions to the translocated native list, and are now feral in the Clarence River well south of their natural range, probably from 'contaminants' in shipments of silver perch from Queensland intended for aquaculture purposes. This is not a large fish but it is aggressive for its size, and is known to breed readily in ponds and slower-moving waters in Singapore, where it has become a popular aquarium fish. Declared noxious in an attempt to prevent further spread, it is now yet another aquatic problem that must be worked around, because it is now too late to do anything about it.

Plate 1

An inland stream may look idyllic while the water is flowing but extremes are the rule across most of Australia, and the animals here have adapted in very different ways. Spangled perch will swim upstream through floodwaters in a search for potential new homes, even on their sides through very shallow water, while in years of drought freshwater crabs seal themselves into deep burrows.

Plate 2

The diversity of wetlands: a coastal waterfall vividly painted with algae, mosses, liverworts and bacteria faces the Southern Ocean; limestone sinkholes on the arid North West Cape trickle water from occasional rainstorms into an underground world of eyeless fishes and shrimps; the buttongrass plains of Tasmania need fire to kill encroaching shrubs, and provide habitat for animals as different as primitive crustaceans and seasonal visitors such as the endangered orange-bellied parrot.

Plate 3

Not long after the first rains an ephemeral pool in a Perth suburb is filled with tadpoles of the quacking froglet among plants which must grow and flower in just a few months including south-western laceplant and narrowleaf water ribbon. Above, rain falls on Eighteen Mile Swamp on Stradbroke Island, which remains at a constant level even though it is drawn upon to supply drinking water to the nearby mainland, and below salt paperbark and black bristlerush have colonised a seasonally flooded roadside easement created by roadworks.

Plate 4

Kangaroo creek, a small coastal stream in southern Queensland, is an unusually rich habitat with at eight species of freshwater fishes including a vivid male form of the crimson-spotted rainbowfish. Two males of the increasingly endangered honey blue-eye spar at left.

Plate 5

The Boyne River near the Tropic of Capricorn in Queensland is at the northern end of the range of two very different reptiles. The common snake-neck turtle does well here among huge schools of rainbowfishes, but is also widespread in the cooler parts of south-eastern Australia. The eastern water dragon climbs and hunts for small terrestrial animals near the shoreline, retreating to the water only when it feels threatened.

Plate 6

Saline lagoons near the mouth of the Murray River are stained orange-pink by red saltpan algae in the warmer months, lending its characteristic red colour to copepods, the elongated native brine shrimp at left, and other crustaceans that feed on it. Many of the fringing plants such as marsh samphire are also common around both inland and coastal saltmarshes.

Plate 7

Mangroves grow where freshwaters meet the sea, and are among the richest aquatic habitats with specialised creatures such as silverstripe mudskippers, a fish that climbs and breathes air. The young of many marine fishes including commercially important species such as mangrove jacks forage over the mudflats at higher tides.

Plate 8

Only swamp she-oaks and river lilies grow along this exposed and apparently barren shoreline of the Myall Lakes in central New South Wales. Linked to the sea in many places, the slightly saline lakes are an important breeding and feeding ground for diverse fishes, prawns and birds. Huge flocks of coots form into ribbons and rafts over kilometers of water as they feed on widgeon grass, and shoals of cormorants will sometimes hunt in tandem with pelicans.

Plate 9

The little galaxias of colder regions of south-eastern Australia breeds in shallow winter floodwaters. In its summer habitat in backwaters (here, near Cardinia Creek from which the species was originally described) it may retreat into marsh yabby burrows in exceptionally dry years.

Plate 10

The multiple roles of plants in wetlands range from submerged species releasing streams of oxygen bubbles to saturate the clear waters of Piccaninnie Ponds in South Australia, while slender cumbungi is an efficient coloniser of exposed wet soils offering a limited kind of habitat to some animals, and the insect-chewed leaves of the sacred lotus in the far north are both food and habitat as the leaf damage indicates.

Plate 11

Crustaceans are generally less mobile than their insect relatives, and are often an indicator of good water quality. A) Glass shrimps are among the most common larger crustaceans of streams and lakes in eastern Australia. B) The scientific name of keelbacked water fleas refers to their distinctive helmet; these are all females and many are carrying eggs in the back of their transparent carapaces. C) Common amphipods are also called scuds or sideswimmers, and are most abundant in relatively still waters with abundant decaying vegetation. D) Yarra spiny crayfish is one of the many species of spiny crayfishes that prefer cooler, well-oxygenated moving waters.

Plate 12

Ferns including bungwahl and climbing maidenhair have colonised the densely shaded floor of a subtropical swamp, providing habitat for the adaptable green treefrog. In the more open waters of a creek just metres away, purple-spotted gudgeons are common but difficult to see despite their apparently vivid colouration.

Plate 13

After a decade of drought Lake Colac is dry by early 2009, with only a trickle from Barongarook Creek maintaining one last, stagnant pool. Carp skeletons mark the level at which the evaporating water became too saline for this unwelcome species, but with the returning rains young spotted minnows and common jollytail (which normally breed in estuaries) appear in schools of many thousands.

Plate 14

Seasonal drought is a regular feature of life in many Australian wetlands, but animals such as this freshwater mussel from the Murray River can survive without water for at least a year. As backwaters and billabongs dry out their trapped inhabitants draw predatory waterbirds to the stranded feast, leaving their tracks on the drying clay, and with the return of the rains short-lived crustaceans and other invertebrates will breed here in uncountable billions.

Plate 15

A creek in the Northern Territory acts as a seasonal pathway linking many deeper pools through the floodplains, where fishes such as the black-banded rainbowfish and seven-spotted archerfish breed during the wet season. Their young spread through surrounding wetlands before the waters recede, but many will be trapped in apparent refuges which dry out, providing food for diverse waterbirds.

Plate 16

A natural lake formed by a landslide in the ranges behind the Great Ocean Road was originally 25 metres deep, a raw sore much like a new farm dam. Now much shallower through accumulation of silt it provides habitat for native predators including large short-finned eels and a breeding colony of platypus, though both must compete for their prey with the introduced brown trout.

Plate 17

Most wetlands around Australia are still suffering the effects of poor management and planning in the past, from floods made worse because of drainage of wetlands surrounding rivers, to the introduction of weeds such as arrowhead. Even now the damage continues – above, a marina being cleared in the acid sulfate soils of an estuary.

Plate 18

The tranquil waters of the Harriet River on Kangaroo Island look pristine, but the dead marron washed up on the foreshore hints of the presence of an invading army of 'native' crayfishes that have changed its ecology forever. Above, the design of this fish passageway on the Murray River is suitable for salmon, but does little for native fishes, while below a badly planned created wetland must be watered to keep plants alive, and has been taken over by the introduced weed drain sedge.

Plate 19

Three of the most destructive feral animals found in wetlands – cane toad, plague minnow and feral pig.

Plate 20

Many valuable wetland habitats go unnoticed by humans. This shallow farm dam sometimes dries out completely, but after winter rains it fills with numerous invertebrates including huge numbers of water fleas, along with more conspicuous invertebrates such as shield shrimps (left) and fairy shrimps (right) which attract predatory waterbirds.

Plate 21

A low-lying, shallowly-flooded corner of a pasture provides a rich habitat for a wide range of unusual invertebrates over many months. Seed shrimps (above) and an unusual black-shelled water flea appear within weeks of the pool reforming, but as it dries out the dominant animals are two species of copepods (below), and several types of insect larvae shortly to turn into adults and fly elsewhere as the pasture grasses reclaim the flooded ground.

Plate 22

Despite its lush appearance this wetland in the Northern Territory is secondary regrowth, with swamp fern scrambling up the trunks of Livistona palms to create varied habitats for many types of smaller animals, both aquatic and terrestrial. Other repaired or improved habitats include these nesting boxes used regularly by chestnut teal (the male standing guard while the female sits on eggs inside), and a weir changed from an obstruction to a fish ladder with the addition of a truckload of rocks.

Plate 23

An unintended habitat: this borrow pit in south-western Australia was mined for gravel to make the adjacent road and has been densely colonised by a wide range of cordrushes around the water's edge. It now acts as a drought refuge for all eight indigenous freshwater fishes found in the area including the enigmatic salamanderfish (photographed in the peaty, tea-coloured water of its preferred habitats), as well as freshwater crayfish and young oblong turtles.

Plate 24

Scenes from the author's swimming dam: pobblebonk (eastern banjo) frogs mating, kangaroos drinking at the end of a hot day, and a green waterspider watches for passing fishes from a waterlily leaf.

Plate 25

This small farm dam has a surprising diversity of self-established indigenous plants, and is an occasional breeding place for up to six dragonfly and damselfly species including the fiery skimmer shown. Few other invertebrates thrive in any numbers here, as dragonfly larvae (mudeyes) are ferocious aquatic predators despite their small size.

Plate 26

Waterbirds which thrive even in the company of humans: the thread that links cattle egrets, clamorous reed-warblers, and purple swamphens is water, though these birds occupy very different niches. Reed warblers weave their nests among the stems of common reed though they aren't a true waterbird, cattle egrets may roost and breed in wetlands but forage in paddocks in the wake of cattle, while swamphens feed on young shoots and plants in wetlands and on the grasses around them.

Plate 27

Nesting sacred ibis in this urban Brisbane wetland will range widely once their young are fully grown, often travelling great distances in mixed flocks with straw-necked ibis, and feeding on grubs and other invertebrates in paddocks. Wood ducks breed in hollow trees near farm dams; here a female orders her well-camouflaged young to freeze for the camera.

Plate 28

During a dry year an unusually large number of swans has gathered in a shallow created wetland fed by urban runoff, but gradually thinned out to a few of the more aggressive pairs by breeding season, making their nests among the thick, fleshy leaves of water ribbon. Other birds such as black-winged stilts fly in to rest in a protected place after feeding over nearby mudflats, while underwater, the largest predators are water beetle larvae, feeding on the abundant planktonic life that has appeared when the wetland refilled after a brief dry period. As they grow larger they will even turn on each other in the absence of other substantial prey.

Plate 29

White-faced and Pacific herons are usually solitary foragers, but here a large group has gathered to feed on trapped tadpoles and larger invertebrates on the shallow, silted-up floor of a senescent farm dam.

Plate 30

Adults of most aquatic insects can fly so they are often among the first colonisers of new dams and wetlands. A) Mosquito wrigglers (larvae) are the most familiar aquatic invertebrates as they appear even in small containers of water on suburban blocks. B) Backswimmers are perhaps the most important and abundant predator on mosquito larvae where fishes are absent. C) Needle bugs are clumsy swimmers but excellent ambush predators, and may appear in large numbers even in new dams. D) Water boatmen prefer mature wetlands where organic matter has been building up for decades.

Plate 31

Few animals live in more extreme environments such as this creek covered with the introduced weed water hyacinth, here covering a subtropical creek so densely that it prevents oxygen from reaching the water below. Only a few tough and tolerant invertebrates with a rich supply of red blood cells to take up oxygen in deficient conditions survive here, including a dark-red aquatic worm (at right) and the bright-orange bloodworm larvae of chironomid midges.

Plate 32

This billabong south of the Murray River is kept at a constant level by a reservoir downstream, allowing a dense carpet of water shield to form among cumbungi and robust water-milfoil. This is also breeding habitat for several native fishes including the blunt-nosed form of the Murray River rainbowfish, southernmost population of a widespread and ecologically important group. Macquarie turtles and freshwater catfish also breed here at times, but frogs are scarce as a result.

Repair

7

Management and restoration

Wetland management can be split into several components, and at the most basic level is a passive process consisting mainly of observation – the path many people would prefer to follow. This is often a risky approach because many of the problems besetting contemporary wetlands can change their dynamics rapidly, and managers need to be prepared to take action at times or put populations of some animals at risk, potentially destabilising the entire ecosystem if these disappear through inaction. Habitat values of a wetland can be degraded by invading weeds or vermin, drainage issues and water quality problems in urban areas, and these have already been discussed in earlier chapters. Later chapters will include information on more specific problems besetting particular species or groups of animals.

In this chapter we are mainly concerned with ways in which restoration (also known as rehabilitation) can be used to correct some of the more obvious problems threatening entire wetlands, with an emphasis on the problems of restoring water levels in drained wetlands, and freedom of passage and shelter to fishes and other aquatic animals in streams. The genetic problems of small populations in isolated wetlands can be critical but are often overlooked, and this aspect is considered in the next chapter along with the last-ditch resort of captive breeding. Management and improvement issues in farm dams and created wetlands are considered in Chapter 9.

Setting goals

A report on wetland rehabilitation in Australia published over a decade ago, and looking at a range of specific publicly funded projects found that 'among the common shortcomings of projects, one of the most disturbing is the apparent rarity of projects with clear and specific objectives' (Streever 1997). Surprisingly little has changed since that time, although an increasing number of projects put much effort into bureaucratic processes that *seem* to be setting goals through elaborate community consultations, rather than in actually tackling the urgent issues.

The resulting draft plans can be agonisingly slow to develop, as anyone who has tried to achieve anything while sitting on a series of committees will know, because all too often these are dominated by a few people who like the sound of their own voices and aren't good at listening to others. Yet we already know very well what is needed to restore or maintain the health of many wetlands, and the first step is defining clear goals in plain English for management, which should also incorporate actions for restoration if necessary.

These are separate things, yet restoration is often a necessary first step ahead of formulating longer term plans for management, especially where it is evident that the health of a particular system is in decline. There is, however, no point in so-called restoration if you don't have a good idea of what a wetland, river or estuary was like in the first place, and although records may be sparse or even nonexistent, historical research is a necessary starting point. Unless you are lucky, this is unlikely to yield anything more than a general idea of the types of plant community that were present, and a list of a few of the more conspicuous animals. You will almost certainly need to supplement this information with a study of comparable but relatively undisturbed wetlands in your region, assuming there are any.

Some goals may seem to be self-evident even without any great depth of research; for example, making it possible for migrating fishes to pass a weir, or blocking drains to attempt to restore a recently drained wetland. How this is best done is usually more complex, however: do you remove the weir or install a fishway, and are there aquatic or semi-aquatic weeds in the drained wetland which will become a problem if not dealt with first? Was the drainage done so long ago that plant communities have already adapted to the new water regime, so that significant types of vegetation will be drowned if past levels are restored?

Goals are often constructed around an increasingly threatened animal (or sometimes plant), which is selected as a flagship species, but this can draw attention away from any wider processes that threaten the entire system. Having a 'personable' animal to give a recognisable face or theme to a restoration project may be useful for publicity purposes, but if it isn't possible to do anything about the causes of its decline (for example, a frog with no immunity to chytrid fungus), this may backfire with its disappearance.

Research and recording

By the time you have completed all possible background research, whether for restoration or long-term management of a wetland area, you will probably have found yourself frustrated by how little of the information you would have liked is actually available. Don't add to the frustrations of *future* generations of wetland managers by keeping everything you see and learn stored only in your head. Pass on what you have already learned in written and photographic form, and keep a record of further changes you observe or make yourself.

Start with a summary of what you have found through historical and biological sources, including where those sources are archived, even if this is just the local library. Adding a few paragraphs on a weekly or monthly basis describing changes observed (or lack of them) doesn't take long either, but you should only record the more interesting or intriguing observations, as few potential readers are likely to want to go through hundreds of pages of minutiae. An annual editing of the record will help keep it concise, cutting away entries that have proven to be of no real interest in retrospect.

Taking photos is an excellent way to record information, and is easy, quick and inexpensive with a digital camera. Downloading into a documents folder with an appropriate record number, caption and date, while backing everything up on a memory stick, only takes a few minutes each time you make a new entry. As the folder grows and it becomes more difficult to find particular entries, splitting it up into further folders on specific subjects will keep it manageable.

Each folder should contain a relevant photographic record that will help make what you write simpler to follow. Some photos will inevitably appear in several different folders as your records grow, and will help to tie together various aspects of an increasingly complex story. Apart from photos of particular events, plants and animals that draw your attention, establish a few photo points to be regularly used to keep track of changes across the entire wetland over the years. Drive in a numbered stake or other permanent marker at each photo point, and when you take photos do them at each and every photo point at the same time. There is no rule about how often reference photos should be taken but I suggest annually at a minimum, better still every few months as photos that turn out to add little of value to the accumulating files can always be thinned out from time to time.

Before making changes

Once upon a time, wetlands were either managed passively or actively but gradually it was realised that passive management, which involves doing nothing at all, is often a contradiction in terms. In the real world, managers have to make decisions at times, whether these turn out to be good or bad, and we don't know enough about many types of wetland or their inhabitants to be able to get everything right all the time. As emphasised throughout this book, the first step to making decisions of this kind is learning what you can about the biology of specific animals and plants of interest, in the context of the overall ecology of any particular wetland.

The books and texts recommended in an appendix to this book will give readers a useful introduction to the various wetland animals and plants, but are only a starting point. In many cases these will provide enough information for general management purposes, with useful keywords for further research as well as a bibliography that will guide you to more specific references, whether regional or for a particular species. A guide to effective location, use and interpretation of information from the internet is also included.

Although the details of what we need to know vary from wetland to wetland and one river system to another, the major overall management issues can be briefly summarised as a matter of knowing or deciding where water levels should be and for how long, and whether we can improve existing habitats whether by restoration or additions. Improvement can be a contentious issue, especially the unsubstantiated belief some people hold that adding islands, nest boxes and other clutter will make it possible to attract and sustain larger populations of favoured animals – see Chapter 9 for discussion of other aspects of this problem.

Undoing drains and fencing

One of the great obsessions of early European settlers in this country was drainage, despite this being the driest liveable continent in the world, and they succeeded at it so well that up to 90% of all wetlands have gone in some areas. The initial damage was usually followed up with grazing, so that more palatable plants still hanging on in lower, wetter areas would die out after a few more years without even the chance to set seed.

Apart from the disappearance of large areas of wetlands and entire types of habitat, there are many types of flow-on effects from drainage including reduced rainfall (the higher moisture levels over forests and extensive wet areas can trigger the fall of rain), and probably in many cases, reduced groundwater recharge. Falling watertables underground are usually attributed to excessive pumping of this resource, but it is also likely that they are also not being recharged as much of the surface water in some areas now runs off almost as quickly as it falls, without soaking in deeper.

If a wetland has been fairly recently drained, say within the last 20 years, there is a real chance that there will still be a viable seedbank of at least some of the original species shallowly buried in the soil and many could make a natural comeback. Seed of many aquatic plants retains its viability for much longer periods if buried in silty soils that remain moist for much of the year, with some species famously able to germinate decades after being buried in such conditions. There are, however, many wetland plants with seed that is much shorter-lived, and these would have to be reintroduced if they don't arrive spontaneously with birds or blown by the wind.

An essential part of encouraging seed regeneration is fencing out of livestock. There are some aquatic plants that will tolerate a certain amount of grazing or trampling, but the longer this continues over the course of a year, the fewer species will survive. Such fences can also be used to keep other vermin out of a re-flooded area including foxes and rabbits, and a simple method of creating a more elaborate fence of this kind was described in the previous chapter.

Even where extensive areas of wetland have been drained a long time ago, there may still be deeper pockets that have survived the drainage with their vegetation largely intact, and these plants may spread if potentially competing weeds are removed before the original water level is restored. In many cases most species of the original wetland plants will be long gone, in which case replanting will be necessary, but this has been discussed in considerable detail in *Planting Wetlands and Dams* and need not be considered here.

Before blocking drains you will need to decide if you are actually restoring habitat values and species diversity, or whether the change will just drown those in what would become deeper areas of the wetland, pushing still more indigenous plants over the brink into local extinction. If the drainage has not been too complete, many of the original plant species will have moved down into lower-lying areas and re-established themselves on relatively bare areas of wetland floor, free of competing plants.

This does not mean they would be able to reverse the process, because the higher ground around the re-flooded area will often be occupied by pasture grasses and other tough and unmanageable plants by now. Many wetland plants such as paperbarks can only move into new places in the form of seed, and while the parent plants will be drowned immediately by a rise in water level, there is little chance that any of their tiny seeds will be able to germinate among the rank growth of weeds which generally develops as soon as grazing animals have been fenced out.

Restoring rivers

Even as European settlers were draining water off the land as fast as possible, they were also blocking off rivers with weirs and dams, and de-snagging the waters in the interest of navigation and faster drainage. It is an unfortunate combination that means at times of heavy rain, more water rushes through rivers because there are fewer wetlands to hold it back (see Plate 17), so floods become worse. The advent of large dams and irrigation systems, especially in inland areas, has also changed the timing and nature of environmental flows as water was often released from inappropriate depths in the wrong seasons.

Even now with improved river management many freshwater fishes still rarely have a chance to breed and some are in danger of disappearing, because the breeding stimuli they have evolved with have been scrambled beyond recognition.

On smaller streams, a barrier 30 to 40 cm high will stop even those few fishes which can jump obstacles on their way upstream, or even endanger them if they clear the barrier but fall onto rock or soil. The barrier need not be solid to be impassable; even a pipe passing under a road culvert with the downstream end suspended a few centimetres above the water represents an intellectual challenge fish have not evolved to cope with.

In more recent times fish ladders are increasingly being constructed to allow free passage past some of the larger barriers, but like many projects implemented by bureaucracy much of the initial effort was misguided. Many of the first (and most expensive) ladders were basically copied from North American designs for salmon, highly evolved and powerful leapers able to make their way upstream past cataracts and waterfalls, as long as they are able to rest occasionally in relatively still pools.

Nothing more different from the sluggish waters of inland Australia could be imagined, and I still remember my shock on seeing one of these brutal constructions attached to a weir on the Murray River (see Plate 18). Looking like a series of concrete prison blocks, it was flushed by a rushing torrent with occasional relatively still (but open and barren) corners for salmon or trout to rest, though these were so wide open to predators that few native fish would have felt comfortable in them.

The only direction most inland fishes could easily have passed along this frightening-looking contrivance was downstream, and they would possibly have been seriously battered by the time they got to the bottom. Remarkably, no-one had thought to improve the chances of a fish successfully negotiating the rapids without being eaten by cormorants by stretching shadecloth across the top – a simple procedure which would cost a tiny fraction of the concrete construction itself.

More recent designs apparently work better as they have been adapted to the biology of the indigenous fishes expected to use them, but these remain in the province of the corporations that construct and maintain the weir systems, and such designs aren't useful or economically viable for smaller-scale projects. Fortunately, the effects of smaller barriers can be undone with much simpler designs and means.

This can be most simply demonstrated with the smallest barriers, the common type of concrete weir up to around a metre high, with a slightly lower apron in the centre over which water tumbles during periods of normal flow. Although the flow rate in such cases is usually small, it effectively creates a small waterfall that few indigenous fishes can pass, but the channelling of the flow into one narrow width can often be undone with as little as a single truckload of large, irregularly sized rocks and small boulders (see Plate 22).

These need to be large enough that they won't shift during floods, but also small enough that they can be manoeuvred downstream by two or three people with the help of crowbars, as the delivery truck is unlikely to be able to approach the weir closely. There is no set pattern to how the rock should be placed; the aim is simply to build up an irregular series of small cataracts up to the lip of the apron, where previously there was just a straight fall.

An arrangement of this kind has no effect on the original purpose of the weir or the pool it creates behind it which may actually provide useful habitat for some fishes and larger crustaceans, as long as they can reach it! If there is enough rock suitably arranged downstream this will create many slower passages through the rushing waters, and relatively still pools and pockets where even the weakest swimmers can rest in a protected place on their way upstream.

Larger weirs will need more rock and more effort to achieve a similar result. Although the visual effect of a rough rock cataract is good and will come to look reasonably natural with time, once barriers reach a certain size they are likely to need specifically designed ladders and cataracts to make the most of both materials and effort. Most of these have not been particularly successful in allowing smaller fish species to pass, although vertical-slot fishways seem to be regarded as the best chance to improve on previous designs at present.

These use a combination of slope and spacing of slots to theoretically allow even fairly weak swimmers upstream, but I have doubts that they adequately address the issue of how far and hard a small fish is likely to push against strong currents. Other useful ideas and designs for fish passages can be found in Chapter 12 of *Innovations in Fish Passage Technology* (1999), and though most of that book is oriented towards salmonid fishes, Chapter 1 also includes some useful ideas that will work if adapted to Australian conditions. Other effects of weirs and dams on water temperature and spawning are discussed in more detail in Chapter 4, along with snag removal and dealing with serious stream erosion.

As with drained wetlands, many riverbanks need to be cleared of introduced weeds such as willows, and replanted with the same types of indigenous riparian vegetation that was cleared in the first place. It is estimated that a belt of vegetation six metres wide absorbs around 90% of the sediment and nutrients (particularly phosphorus) which would wash directly into a stream with bare banks. Similarly, fencing is essential to keep livestock out, and although fox- and rabbit-proof fences would be desirable along streams, these also restrict access for smaller native animals, and are difficult to effectively patrol and maintain over long distances as compared to the often much shorter perimeter of other types of wetland.

Improvement – a meaningless term?

Rehabilitation of an obviously degraded wetland or stream is one thing, as in these cases we are consciously trying to return to an earlier and more natural condition. Improving habitat values is another, and almost invariably falls into the trap of visual appeal, as opposed to what the animals that are supposed to benefit will actually gain. Some types of improvement may work to a degree in a large and open wetland, yet their value becomes increasingly doubtful as they are applied on an increasingly small scale; islands and extended shallows are a good example of this effect, and are discussed in Chapter 9.

The central question in all of these cases is just exactly what is being 'improved', and this can be seen most clearly by exaggerating the numerical 'benefits' which should accrue. Let us use a series of improvements aimed at increasing numbers of black ducks as an example, and assume that as a result of these the number of ducks has been increased a hundredfold. I doubt there is anyone who would not be instantly horrified at this notion, because water quality would be the first casualty. The only survivors underwater would be worms and bloodworm larvae (see Plate 31), which are able to survive in low-oxygen environments, while the shoreline would rapidly be undermined by the constant dabbling activities of the ducks.

Most plants would disappear apart from cumbungi, common reed and some of the larger sedges and rushes, which would thrive so well in such a high-nutrient environment that they would soon cover all of the wetland that wasn't too deep for them – ultimately driving the ducks away. In reality, this scenario is beyond the realms of possibility, but begs the question of what 'improvement' really means, and at what stage it becomes a caricature of the natural world. I would argue that any change that shifts an established balance in any healthy wetland has little to recommend it. Even if water quality is only 10% worse as a result of artificially enhanced numbers of ducks or any other larger animal, this would already be a first step towards reducing diversity by weakening or even killing off more sensitive animals and plants.

A further problem is that such improvements are almost universally perceived in terms of increased numbers of birds, though even the most tolerant species will not linger in a contaminated and potentially lethal environment. Yet some groups are trying to improve habitat values along these lines, for example by adding nesting boxes that favour the breeding of selected species. The nest boxes installed by duck-hunting clubs in particular are designed for the benefit of species it is legal to hunt, and not any others, with no consideration for the impact a numerical increase in such already common birds would have on the overall environment or on other birds.

Even then the job of installing nest boxes is often done amateurishly, without consideration of the needs of the birds themselves. In some wetlands the only birds that breed in the boxes are swallows, because there is no suitable nesting material within kilometres that such clumsy fliers as ducks can easily carry any distance. The addition of straw pressed into a rough nest shape in each box will encourage ducks as they then only need to add breast feathers as an insulating layer for their eggs (see Plate 22).

Nesting boxes have become a common and accepted feature in many urban wetlands, which has blinded us to their ugliness, and the way they destroy any feeling of wildness and the beauty of open water. I can imagine the outcry if tyres were also thrown in to increase crayfish numbers, and perhaps bells attached to ropes that ducks can pull when they want to be fed. It would be a bit like filling northern hemisphere forests with kennels as winter dens for wolves.

In the final analysis, changes to favour one species over any others in a wetland are a dangerous thing, which can backfire in many ways. All such changes are subject to our personal biases including attracting unnatural numbers of birds, the most common problem of this kind, and are more likely to reduce diversity overall than increase it. Nor will all species that die out as a result of excessive predator pressure, deteriorating water quality, or any other more subtle and unpredictable factors necessarily return if the problem is corrected later.

Permits

To avoid unnecessary duplication in many of the chapters which follow, the issue of permits is dealt with briefly here. Permits for all aspects of dam and wetland creation have been discussed in *Planting Wetlands and Dams* (2009). If you are planning to introduce any kind of animal into a wetland, the first step is to investigate state laws regarding the species you have in mind, and whether there are any restrictions on keeping them captive, or moving them around.

Although the relevant laws vary from State to State, frogs usually may not be shifted from one area to another, partly because of the significant risk of spreading fungal and other diseases to unaffected populations, but also because some species have been carelessly introduced into places they don't belong and have thrived there. In general, the policy seems to be that if indigenous frogs aren't going to arrive in a wetland or dam under their own steam (see Plate 24), they either don't belong there and probably won't thrive, but even if they thrive they aren't welcome additions to the local ecology.

Fishes that are regarded as threatened in any way are also mostly rigorously protected, though not always from the actions of other government departments, not least because shifting them randomly between one wetland and another is often a dismal failure for various reasons discussed in Chapter 11. Reptiles, birds and mammals are not usually introduced into wetlands as they will find their own way if they are already found in the area, but even if there is good reason to attempt an introduction, a permit *must* be sought first.

Permits are not automatically given out, and in the case of many species there may not even be a protocol in place for creating one. Usually, if a good case is made for stocking or reintroducing an aquatic animal to a wetland it will be considered, providing that the biology and habitat needs of that species have been fully considered. And as has been emphasised throughout this book that requires research – as detailed and up to date as possible.

8

Populations, genes and captive breeding

This chapter looks at the genetic aspects of managing animal populations in a wetland, including how large a population must be to remain viable in the long term, and the thorny issue of whether it can be rebuilt through a captive breeding program if numbers fall to an unstable level. Plants are not considered here as their biology is different enough that a different set of considerations applies, but also because most replanting is done with plant material collected and propagated directly from wild populations.

Population size and genetic problems

Whatever type of animal is being considered in later chapters, an issue that must affect all of them, and also the planning needed for long-term management of a wetland for particular populations, is sustainable population size. This can best be described by the '50 to 500 rule', where at the lower end of the scale a population of 50 or so animals of a single species may *theoretically* be genetically diverse enough to scrape by from one generation to another, as long as no major disaster such as fire, extreme drought or invasion by a serious predator or competitor occurs.

In the real world, there is no guarantee of such stability, and even relatively minor mishaps may gradually eat away at variability in a small population over many generations; for example, if only half a dozen of them carried a particular gene, and all of these just happened to die out simultaneously for any reason. It need not be that these genes are unfavourable; in a small enough population such an effect can happen every now and then by chance alone.

Over long periods of time, one event after another of this kind will happen through the laws of chance if nothing else, so the range of genes represented in a small population diverges from those more typical of a larger and more stable one – this is called genetic drift. The process can be accelerated by a related phenomenon, where after some disastrous event only a small number of survivors remain to breed the next generation, and more genes are lost as the population is squeezed through a 'bottleneck'.

All of these ways of losing genes lead eventually to inbreeding, where the genes of most individuals become increasingly similar, with the addition of faulty copies that would often have been masked or diluted by the correct copies in larger populations. At this point in time it may only be possible to salvage the remaining diversity in that small population by 'out-crossing' with another population that has a different set of genes, but common sense tells us it is better not to let a population dwindle into such a state in the first place, and to do everything possible to keep it in the hundreds at all times.

Not as obvious is the effect of the natural lifespan of a species – the shorter the generation time, the sooner all such effects are likely to strike, as an animal with a lifespan of only a few months or a year will obviously go through a lot more generations in a decade than another type of animal that lives 10 or 20 years. On the other hand, in a relatively long-lived species there may only be a small group or even just a dominant pair of mature animals breeding in a group of 50 so the population is functionally much smaller than the total number of animals, and inbreeding is also likely to happen much more quickly.

For these reasons, a minimal sustainable population size of any one species of animal is usually regarded as being in the order of around 500 individuals. This is not a figure set in concrete but a guideline, although it does make it obvious that introducing a pair of frogs or fish to a wetland where there is little chance of contact with others of their species is not the ideal way to try to establish a viable breeding population. For these reasons, an isolated frog pond in a suburban backyard is not an adequate habitat, as it is unlikely to support more than several breeding animals at any one time. On the other hand, 100 or so frog ponds on the same suburban block, within an easy frog's march of each other, *could* potentially constitute a sustainable, long-term habitat when considered together. For the same reason garden ponds won't support viable populations of even the smallest fishes in the long term, and anyone planning to develop breeding colonies will need to swap around some of their stock with other like-minded people every few years to maintain a wider gene pool. These considerations particularly affect animals which aren't able to commute to nearby waters, mainly fishes and to a lesser degree frogs, as well as various invertebrates such as snails which can't get around much on land. It does not affect flying animals including most insects and all wetland birds, as these are able to move on if conditions no longer suit them, so any wetland they are found in can be considered to be just a single part of the mosaic of habitats they occupy. Nor is it a problem for most microscopic creatures, as a wetland or pond would have to be very small indeed to not be able to sustain a population of thousands, and also many of these have their own ways of dispersing between wetlands such as hitching a ride with birds.

Captive breeding

Many aquatic animals are isolated in particular wetlands, and if conditions there deteriorate or a competing species invades, their population may drop dramatically to an unsustainable level in a matter of generations. One apparently common sense response to this is usually along the lines of removing the survivors from harm's way, establishing and breeding them in a safe environment, and reintroducing them later.

This is called captive breeding, and while there have been a handful of successes over the past century, such programs are usually dogged by poor or (even worse) ad hoc planning, lack of clearly defined goals, and often an apparent inability to learn from the experiences of comparable programs which have also failed. Even as a last resort, captive breeding is often left until inbreeding depression (where the effects of increasing genetic problems are already obviously affecting breeding success) has obviously set in, so remedial measures are less likely to succeed.

Readers familiar with my captive breeding programs for various native fishes, several of which are included in *The Action Plan for Australian Freshwater Fishes* (1993), and some frogs in conditions as closely replicating those in the wild as possible, may be surprised to find that I have increasingly come to agree with the critics. After three decades of observing the problems involved, I have come to believe that captive breeding can only work as a short-term emergency measure, if at all.

Proponents of captive breeding tend to concentrate on the logistics of breeding in captivity, and put far less thought into reintroduction or how the habitat that has been degraded could be restored. If a captive breeding program does not include a plan to understand and counteract the problems threatening to wipe out the population in the wild, and to correct these as quickly as possible so that the species can be reintroduced, it will rapidly dwindle into a zoo situation where each generation will become less like its wild ancestors.

Among other costs there will inevitably be some selection for heritable traits that may be desirable in captivity, yet reduce the fitness of the captive-bred animals for a return to the wild. The most obvious of these is willingness (or even just the ability) to breed in unnatural conditions. Of the various types of animals successfully bred in captivity as ornamental animals, there are more types of fishes being regularly spawned and raised for sale worldwide than any other type of animal. Many of these have only been domesticated a few generations ago, and some are still being collected from the wild.

In the process of adapting these to captivity over the past century, it has become common knowledge among fish-keepers that wild individuals are usually much more difficult to spawn than those already captive-bred, even in *apparently* ideal conditions. These problems may stem from a variety of causes, such as incompatibility between wild animals with very precise ideas of what they want in a mating partner, advanced age (not always easy to recognise), and inability to adapt to a set of conditions that humans may imagine are close to those in the wild, but are not recognised as such by the prospective breeders.

Foods (and the hidden nutrients they may contain – see Chapter 11 for discussion of some aspects of this problem) probably figure as a significant part of the problem; for example, zoo populations of growling grass frogs turn a uniform leaden-brown even when fed an abundant diet of captive-bred crickets. Using ultra-violet light on the frogs doesn't have any effect; it is something lacking in the diet, suggesting that the crickets themselves aren't carrying the full range of the nutrients that would be available to them in the natural world.

After successful breeding, future generations of captive-bred animals become increasingly easy to breed even under conditions that are clearly different from those under which they evolved. Even the foods supplied to captive bred animals are usually quite different from those they must recognise and be able to hunt if they are to survive in the wild, so that after a few generations they may become incapable of feeding themselves if released.

Restoration of the habitat itself to something approaching the state it was in before population crashes is also essential but has generally been neglected, sometimes on the grounds that it may have improved by itself, or that it is simply too expensive or difficult a problem to deal with. Thus we have fairly elaborate breeding set-ups used to produced Pedder galaxias for reintroduction into waters still full of the trout and climbing galaxias that have become both its predators and competitors, or randomly bred mixes of captive-bred purple-spotted gudgeons (see Plate 12) dumped into unaltered streams they had long ago disappeared from, as discussed in Chapter 12.

Even where something is planned to be done about the state of a degraded wetland or stream, this is usually focused on a 'flagship' or 'iconic' species. As a result other aspects of the threatening processes may be conveniently ignored when it is often the ecology of the wetland as a whole that is under threat. To quote Horwitz (see p. 80, Serena 1994) 'there is little attempt to reconcile the recovery of high profile species with that of the lesser known ones'.

Eacham rainbowfish – why planning matters

The Eacham rainbowfish was originally described from a population found in a small crater lake on the Atherton tablelands near Cairns, but had disappeared from this locality by 1987. The reasons were obvious. Anglers had introduced several predatory native fishes, of which mouth almighty and banded grunter can certainly catch adult rainbowfishes, while archerfish may feed on smaller ones. The rainbowfish had lost all awareness of other fishes as potential predators over the many generations it had been breeding in the lake with no predators present, much as aquaculture fishes in protected environments do.

As the first indigenous freshwater fish to apparently become extinct it attracted some media attention, but the Australia New Guinea Fishes Association (ANGFA) located two small, aquarium-bred populations and decided to use these as the basis of a captive breeding program. I was the only member who voiced any concerns that the two small groups to be used had already been through some severe genetic bottlenecks (including probably excessive culling of supposedly undersized fry – see Chapter 12), and that there was no prospect of returning the fishes to the lake without complete removal of the predators. These were not good starting points for a captive breeding program which was being set up as an environmental flagship for the society, as aquarium breeding is a guaranteed way to favour only the fastest growing individuals.

Within several years it was realised that there were other populations of this fish in nearby streams, apparently under no threat at the time. This meant that the handful of aquarium fishes from the lake population were no longer the last of their kind, and the news also came at a time when reports of increasing numbers of problems with introduced strains of piscine tuberculosis were appearing in ANGFA's newsletters. Despite its potential to become a vector for this disease and possibly others imported through the aquarium trade (many aquarists mix fishes from many sources) into relatively pristine waters, several ill-conceived attempts were made to reintroduce it to the lake. As nothing had been done about the introduced predators, the released fishes predictably disappeared within a short time.

For all the publicity generated over the years, the disorganised nature of the entire 'program' for this particular population meant that there had never been any coordinated approach to keeping and breeding a reasonably sized population in conditions remotely approaching those under which it evolved. Nor was there any attempt made to keep track of who was maintaining it, or where, or under what conditions. Even the population held by Taronga Park Zoo, the sole public institution supposedly maintaining it *in perpetuum*, died out after being relocated to the education section of the zoo when the aquarium was closed down.

Recently, in an attempt to bring some order to the miscellaneous collections of fishes kept by ANGFA members, it was found that the sole remaining population of the lake form of the Eacham rainbowfish was a few elderly fish in a suburban Sydney garden pond. These have produced very few fertile eggs at the most recent report, and attempts to breed from them have produced precisely three offspring.

Passed through many different types of bottleneck, possibly carrying exotic diseases they have themselves survived and become immune to, these last pathetic survivors must be almost totally lacking in genetic variation.

To all intents and purposes the lake form of the Eacham rainbowfish, whose rescue from extinction had been trumpeted to the world two decades ago, is now a captive-bred rarity with no prospect whatsoever of being returned to anything resembling its natural environment. And in the meanwhile, 'native' sooty grunters introduced into the Johnstone River have now started appearing in its tributaries including Dirrin Creek, heralding a new threat to the remaining wild populations of this rainbowfish.

9

Created wetlands and dams

Created wetlands come in many different varieties, but can be roughly divided into wetlands intended to imitate at least some aspects of natural waters and ecosystems, and dams which are primarily a way of storing water but can also be modified or improved to provide some habitat functions as well. For all of the good intentions underlying created wetlands, very few have a clearly defined plan or include any provision for monitoring the achievements (or failures) of those who design them. By contrast, the humble farm dam can only be improved regardless of its original limitations as habitat.

Created wetlands

Very few created (also called constructed) wetlands have much value as habitat for anything but the most common and adaptable animals, mainly because they are designed and planted by people who have little or no background in aquatic biology, and lack any clear or defined goals beyond water treatment and a pleasing appearance. Habitat is always accorded a token importance on the plans, but without any indication of just what it might be for, or any recommendation as to assessing what its value.

This is the outcome of a curious phenomenon – the coordinators laying out the grand plan for nearly all created wetlands are mostly landscape architects. I have yet to hear of any Government authority employing a landscape architect to rehabilitate or recreate a terrestrial ecology, yet they continue to be asked to design wetlands as ecosystems, and in some cases win awards for their work because those who judge such things are also usually landscapers. As long as the wetland looks pretty and also attracts ducks from time to time, it is regarded as a complete success.

An attractive appearance is fine and is of considerable concern in urban developments. It is the pretense that such wetlands also create rich habitats which is objectionable, when urban development is the primary cause of loss of diversity in a wide range of ecosystems around cities *including* wetlands. The one ecologically positive thing that most created wetlands do a reasonable job of is water treatment, because the limited range of plants likely to survive the semi-toxic soils and waters of newly created wetlands are invariably colonisers that will also use up a wide range of nutrients.

A further problem in most created wetlands is the planting lists, often a mish-mash prepared by people who seem to have no idea of the guilds found in nature, or their tolerances to each other. I call this the 'Mallee with treeferns' approach – the diversity of wetlands and their associated plants has already been emphasised, and random combinations of aquatic and water's edge plants are often as visually and ecologically implausible as planting treeferns among mallee scrub.

Even when the planting lists are reasonably appropriate, they don't address a range of other problems often seen in created wetlands, particularly the issue of provenance. There is not a single project of this kind that doesn't emphasise the importance of using locally sourced plants, but the short lead-times for propagation, and the indifference of some of the larger companies which do the planting means that the plants actually used are whatever is available at the time. In turn, that means whatever the largest wetland nurseries have on hand, and as these are all in the largest cities, the range of species and provenances is invariably limited.

As a result of the recent fashion for water treatment wetlands to deal with runoff from new housing developments, the plantings themselves may be still further constrained by bureaucratic processes bearing no relationship to the physical realities of a site. Among the stupidities I have seen in recent years, an increasing number of dry wetlands have been planted because this was a requirement before a development could proceed, and as a result watering systems have had to be installed in a futile attempt to keep the plantings alive. Only a limited range of wetland plants will thrive under such conditions, and not surprisingly many of them are introduced weeds (see Plate 18)!

Even ignoring good intentions and the dubious question of how well they have been put into practice, there is still no evidence that it is possible to replicate natural wetlands at will, and never has been. Despite this, the concept of 'no net loss' has been used to justify drainage of natural wetlands in inconvenient places, replacing them with the depauperate communities of created wetlands. We have already caused enough damage to the natural world, and conservation or even restoration of natural communities left in anything like their original condition should have priority over the creation of new wetlands.

Some readers will find it strange that after 30 years of working with wetland plants, and developing Australia's first specialist indigenous wetland nursery, I should condemn the overall quality of the outcomes. Yet my interest as a zoologist began with plants in the context of the habitat needs of particular animals, and for all the attention to visual appearance and lip-service to habitat values, at the present time few created wetlands are anything more than a pleasant place to walk a dog. They can be improved upon considerably with more attention to design instead of repetition of clichés and platitudes, as has been discussed in some detail in *Planting Wetlands and Dams* (2009).

Farm dams

There are possibly millions of farm dams in Australia, ranging from little more than small pools to enormous water supply dams that provide a limited type of refuge for waterbirds in times of drought. Although their habitat values are limited in most cases, they are already there, they (usually) hold water, and their attractiveness to a wide range of wetland animals and as a sheltered source of drinking water for terrestrial animals (see Plate 24) can only be improved upon in most cases, though some already have a rich and diverse fauna which goes largely unnoticed.

Kept clear of encroaching vegetation by cattle and fertilised by their droppings, these unfamiliar habitats may support diverse small crustaceans, and some not so small. If they dry out regularly, they will almost certainly be home to millions of water fleas, copepods and other microscopic crustaceans that thrive in their greatest numbers where predators are regularly killed off by drought. Some are not so microscopic – shield shrimps and fairy shrimps (see Plate 20) can reach the length of your little finger, large enough so that as their dam dries out they may attract small flocks of flying predators including terns.

Other crustaceans from temporary waters may be even less common, though less conspicuous. In the temporary pools surrounding shallow farm dams, I have several times stumbled upon small and unusual beasties that have only been found occasionally before, so that it took some time and effort to have them identified. This does not mean that they are necessarily rare (see Chapter 11) but is a reflection of how overlooked the inhabitants of such miniature habitats are.

Before considering ways to improve a dam as habitat for other species, it is a good idea to actually find out what lives there already. Larger animals including fishes and frogs may use or even breed in seasonally flooded areas around dams, and even in water-filled drains which are the remnants of formerly more extensive wetlands. The most barren-looking ditch may provide significant habitat for some inconspicuous yet interesting animals, so don't rush into making other plans for it that may destroy its existing values before you even learn about them.

Although excluding livestock (particularly cattle) is recommended as one of the most basic improvements for most types of wildlife, ironically it is livestock that have made many dams much more like natural wetlands than those which are deliberately created. The erosion they cause along with their manure forms a fine bed of nutrient-rich silt which is far more like wetland soils than any recently flooded topsoil, which can form a fairly toxic brew that only a limited range of plants and animals will tolerate for the first few months.

For new farm dams, a case could be made to allow livestock access for two or three years to speed the maturing of the biological system, before fencing to reduce the rate at which silt accumulates, as too deep a bed of silt is almost as undesirable as raw clay and will restrict the variety of aquatic animals which will thrive. This would also spread the cost of setting up a new dam that is intended to provide habitat as well as drinking water for farm animals, postponing the cost of fencing until a later date.

Improving farm dams

Most farm dams do little more than provide muddy water for a few cows, while breeding an interesting range of waterborne diseases and parasites. It is not surprising that people are reluctant to swim in (let alone drink from) dams where cattle wade regularly, but why would anyone think it is good for the cattle themselves? Some figures suggest that livestock drinking clean water may grow up to 40 % faster than their unhealthy siblings that get to drink from the very water they excrete in. Once fenced and planted to improve habitat values, with a siphon line to feed clean water to a stock trough, a farm dam can become an asset both aesthetically and in terms of production.

The success of a dam as habitat is usually measured in terms of abundance of bird life, and it is often assumed that crowding in a number of dubious additional features and a wide variety of plants will attract more birds. Yet all the evidence suggests that the number and diversity of birds is a function of total area more than any single other factor, and that the larger a dam is, the more it looks like a 'real' wetland to most waterbirds (see Plate 29). Other improvements recommended for fishes, frogs and other species may be even more ad hoc; for example submerged logs that are supposed to appeal to a wide range of fishes.

The most popular plantings from a landcare perspective consist mainly of trees, which may be crowded in around even the smallest dams. These will ultimately cut off flight paths for birds, as well as light and warmth to the water itself. The most dramatic example I have seen was two virtually identical dams on a single property less than 200 metres apart, one closely surrounded by young trees already several metres tall, while the other was other still unplanted.

The owner of this property had already noticed that birds were avoiding the planted dam, yet were still happy to use the apparently barren dam, and the reason for this was clear if you waded into the planted dam. To a bird, the rising walls of trees must have made it seem like a deep, steeply sided pit with a little water at the bottom, and the steep 'sides' combined with the windbreak effect meant that extra effort was required to fly out. For most birds, lift-off is much easier over a longer distance, preferably into a breeze that helps reduce the work of getting airborne.

The most obvious feature of natural wetlands that attract large numbers of birds, without even considering the plant species present, is their openness. In part, this is a function of size, though some smaller farm dams with particular suites of low-growing plants and good air movement also attract birds. Even grebes will breed in an extensive farm dam with an adequate supply of fish or tadpoles and suitable reed beds for nesting material, while in smaller dams predatory birds are more likely to be visitors rather than residents due to the limited food supply available.

As most types of waterbirds feed in relatively shallow waters, the most dramatic improvement you can make to an existing dam is adding an extensive shallow planting area along one side, to form a new wetland in its own right. If appropriately planted, this can encourage a range of new arrivals from diverse invertebrates to nesting birds. Such additions should slope gently down towards the dam itself, although they can also be done as a sort of artificial billabong adjoining the dam, and only connected to it during times of the highest water levels.

A shallow, graded area like this will allow a wider variety of plants to grow than the steeper sides of the dam itself. The beauty of a properly designed combination of new wetland and old dam is that you can use the dam much as before, including for irrigation which is usually only needed once most water animals and birds using the wetland have already completed their natural breeding cycles. Selecting plant species for various rates of drawdown and evaporation rate is discussed in *Planting Wetlands and Dams*.

Use natural wetlands of the same depth in your area as a guide to planting. A wetland dominated by rushes which usually grow in situations where they are high and dry within a month won't attract nesting birds which need water around their nests for two or three months. However, it will attract those which forage over drying mud, and even if their nests are somewhere else in some other wetland, you will still have increased the feeding habitat available to them. On the other hand, a wetland area dominated by plants which grow mainly where water will stand for many months gives different signals to potential breeders, which must be able to recognise suitable breeding places at a glance.

Islands and other pipe dreams

Islands are popularly seen as the best way to improve habitat values in a farm dam, mainly because it is assumed that they will attract resting or nesting birds, even though the birds themselves may become a liability in smaller dams if they are present long enough to affect water quality. The benefits of island building even in the smallest dams became something of a superstition through the 1980s and 1990s, yet surprisingly few such islands have actually been constructed to the single, unvarying formula copied from one book to another, and these rarely attract birds at all.

The best results were supposedly with number of smaller flat-topped islands that look like a little volcano when exposed by falling water levels, yet it is rare to see anything like this in nature and requires earthmoving on a monumental scale, leaving only relatively shallow areas between the multifold islands so the wetland will dry out very rapidly. Logs were supposed to be strewn on the islands to provide nesting hollows, even though trees are rarely found near small islands in natural wetlands.

Floating islands were an even worse idea, and those that have not sunk long ago become unappealing skeletons of pipes and floats that eventually will no longer even hold soil to hide their inherently repellent structure. Pipes were so popular that it was even recommended they should be scattered underwater as fish-breeding habitat, leaving the wetland looking remarkably like a tip when it dries out. Replacing these with logs isn't much of an improvement as trees are only usually found in large numbers along rivers rather than in more open wetlands.

Studies in the USA suggest an island must be at least 170 metres from the nearest shore to discourage foxes from swimming out to them, and that these predators actually *like* to den on larger islands closer to shore as a safe place to rear their cubs. Although not all species of native waterbirds have had experience with foxes until the last 200 years, they are all fairly closely related to or have evolved from species that do, and an instinct to not nest on small islands close to shore should have been deeply ingrained into their common ancestors over millions of years.

The problem with islands in a small dam is that they are an anomaly, steep-sloped (even if the top is flat) and completely unlike anything you would see in a natural situation, so birds probably just don't recognise them as a desirable place to be. The more extensive a wetland is and the larger the islands in it, the more likely it is that birds will use them. Similarly, more species and greater numbers of waterbirds will use a larger dam with more extensive shallows. This is probably a direct reflection of how much food is actually available, so small and shallow dams with tiny islands are unlikely to draw more species or greater numbers than deeper ones of the same area that hold water for longer.

The evidence all suggests that larger, relatively treeless, open dams with extensive shallows are preferred by most species of waterbirds, whether they are predators or herbivores. In other words the less volume of water a dam holds in comparison to its area, the more attractive it will be to a greater range of birds. Apart from being shallow overall and relatively expensive to construct, such dams are unsuitable for irrigation purposes, as pumping from them will combine with evaporation to drop water levels so rapidly that it may be impossible to establish plants, and nesting birds will be left high and dry before their young are fledged.

Interactions

Birds are not the only inhabitants of farm dams, though some will inevitably visit and perhaps even stay for a while if there are enough of their preferred foods to draw them. This is a point that doesn't seem to be considered when talking about the quality of habitat – that there is also an issue of quantity. If a dam has a resident population of frogs which produce (say) 5000 tadpoles a year, and a cormorant or a heron needs 50 large tadpoles a day, then even if the tadpoles just lay around passively waiting to be eaten the available food supply is only enough to support that solitary cormorant for 100 days.

The reality is more complex, as until the tadpoles reach a certain size it is not likely to be worth the predator's time and effort to hunt them. Similarly, once it has eaten enough of them it will become too much of an effort to find the increasingly large and wary survivors – in fisheries parlance 'commercial extinction', or what used to be called the law of diminishing returns. This process also selects for the more wary tadpoles which is good for future generations, except in dams that dry out completely so herons have easy access to their prey (see Plate 29). It is also part of the reason predator-free backyard ponds are unlikely to contribute much in the way of useful genes to wild populations of frogs in nearby wetlands. It doesn't matter that a pair of herons or cormorants couldn't possibly raise their annual brood on the proceeds of one smallish dam, because they will visit a number of dams on their daily hunt, and a shorter hunting period in each dam also keeps their prey from becoming too wary too soon. A little pied cormorant (presumably one of a pair) that visits my swimming dam every few days as part of its rounds during winter and spring also does the rounds of around 20 other dams within a few kilometres.

The number of visits by any one species or individual bird also has a regulatory effect on what the birds do to water quality, because everything they eat is also later excreted. Nitrogenous wastes are a normal part of aquatic cycles (see the next chapter), but an overload of bird droppings will damage the health of all animals that breathe through their gills, whether these are insects, fishes or tadpoles. Too many birds visiting a small dam in too short a time could potentially kill many of the animals in it with their wastes, but spread over many months their droppings help maintain fertility.

For this reason even the most obsessed bird lover should not try to attract more birds to a dam with supplementary feeding of any kind, than it can realistically sustain. The reason so many urban wetlands are notorious for water quality problems is that there are far more birds present than is ecologically desirable, usually ducks attracted by handouts of bread from well-meaning citizens, who are often among the first to complain about the resulting smell and blooms of cyanobacteria!

Birds have been used in this chapter to make some significant points about the confined nature of a dam environment because they are both familiar and popular, yet their ability to come and go overnight in large numbers can make them something of a wild card among wetland animals. Many other animals are smaller and more closely attuned to life in relatively closed aquatic ecosystems including dams, and many of these can even form self-sustaining breeding populations in the right environment.

Yet even this can change depending on the other species present, so a smallish dam that will comfortably sustain a breeding population of 500 pygmy perch will become a very different environment if a predator that feeds on pygmy perch is introduced. And a dam large enough to sustain a breeding population of that predator must also include enough planted area to give shelter to the prey, so that it can breed the thousands of young needed to keep the predator fed. It is all a matter of common sense and basic arithmetic, and in the case of most dams the fundamental message is, think small.

Plants and animals

10

Native plants and habitat

It will already have become obvious from earlier chapters that although plants and habitat are often thought of as being much the same thing, many habitats are characterised by the *absence* of plants, and many animals are not particularly fussy about the plant species present. This is elaborated on in more detail in the next few chapters, where the specific needs or preferences of various groups of animals are discussed in more detail. The basic needs of plants including soils, water conditions, light and interactions have been discussed in *Planting Wetlands and Dams* (2009); this chapter focuses primarily on the habitat values of plants as food, shelter and nesting material, as well as considering the habitat needs of some less common plants themselves.

What defines a plant for our purposes is the presence of green chloroplasts that allow an organism to photosynthesise, using sunlight, carbon dioxide and water to fix energy into more complex compounds, and these are directly or indirectly the source of energy for nearly all other living things on earth. The main exceptions are bacteria and some other microorganisms that live in the earth's crust and use heat energy, and animal communities around deep sea vents which are also dependent on the energy-fixing abilities of bacteria.

Herbivorous animals feed upon plants in one form or another, carnivores feed upon herbivores, and anything that dies and all wastes that are excreted are fed upon by other organisms including fungi and bacteria, to extract what is left of their energy content. In turn, this releases nutrients which are then recycled by plants, and incorporated into increasingly complex compounds. Ecology is essentially the science of what eats what, and how the energy is transmitted from plants to diverse animals and other living things.

Algae and cyanobacteria

Apart from the more conspicuous plants that dominate many wetlands, there are also many smaller species which can be significant sources of food, and in the case of algae may also appear in large quantities in response to high nutrient levels. Algae have been around much longer than most other wetland plants without apparently changing much, and although this suggests they are primitive it actually means they have had a long time to become very good at what they do.

Apart from anything else, an increasing number of species are known to produce allelopathic chemicals, designed to deter or even kill competing species, and to some degree animals as well. These are part of the arsenal of many cyanobacteria (blue-green 'algae') especially, which is why a bloom of some of the more toxic species of these makes water unfit to drink, or even to come in contact with in extreme cases.

Algal blooms can appear in a very short period of time, disappearing literally overnight when conditions are no longer ideal for them, and this is probably the origin of the myth that barley straw will get rid of cyanobacteria. It is more likely that an algal bloom coincidentally disappeared at the same time as straw was added to the water, probably as a way of encouraging plankton in an aquaculture situation – ironically, it does this by promoting the growth of single-celled algae! In my own experience and from other reports, an excessive use of straw of *any* kind is more likely to *cause* blooms than prevent them.

Phytoplankton are mostly single-celled algae as well as some more animal-like organisms that just happen to be green and photosynthesise, and are present in all but the clearest waters. Where they are present in large numbers (usually in waters with a high nutrient content), the water turns greenish, and in extreme cases such as crowded duck ponds becomes a thick and soupy green. In turn, phytoplankton are fed upon by zooplankton such as water fleas and copepods, discussed in the next chapter.

Other types of algae form submerged or floating mats that may be made up of one or several species, or attach themselves to rocks and other solid surfaces underwater. In garden ponds such growth is regarded as unsightly, but it usually only appears in large quantities where nutrient levels are fairly high and there isn't enough plant growth of other species to keep up. Apart from improving water quality, many such algae are also significant foods for tadpoles and some smaller organisms, as well as providing shelter for a range of tiny organisms that may be eaten with the algae themselves.

In African lakes, algal mats with a significant animal component are called *aufwuchs*, and are grazed upon by diverse herbivorous cichlid fishes, many of which would not thrive on a diet of the algae alone. It is possible that the animal component is also important to at least some species of tadpoles, though the long gut needed to digest and assimilate algae suggests that animal protein is only a minor part of their diet at best.

Major groups of habitat plants

The most conspicuous species in vegetated wetlands are nearly all flowering plants, including grasses and sedges with wind-pollinated flowers that are so small and obscure they are not generally recognised as such. Not all plants are equal as habitat, and some of the most common types attract only a limited range of animal species. Common reed is the best known example, with tough bamboo-like stalks and papery leaves that few animals will eat, forming stands that may cover many hectares.

The middle of a dense reed stand seems lifeless on a still day, as even reed-warblers (which tie together several canes to support their nests – see Plate 26) are found mainly around the edges, closer to other habitats where they actually feed. Cumbungi stands aren't quite as lifeless as the young shoots attract purple swamphens (see Plate 26), which may also build their nests near the fringes, though they prefer more open situations if given a choice. Water ribbons are more favoured both as food (as discussed below) and for nesting sites.

Some aquatic and semi-aquatic grasses form extensive stands that may attract birds when they are seeding, but sedges are more important in wetlands in ecological terms, both as shelter and as nest sites. Hollow-stemmed species growing in deeper waters such as some twigrushes and spikerushes are used to make floating nests by some ducks, anchored to the bottom by the stalks, yet these floating nests can move up and down to some degree with changing water levels. Tussock-forming sedges and also rushes offer a different kind of habitat, as they tend to space themselves fairly evenly once mature, leaving shaded and protected tunnels between each clump as shelter for frogs and smaller mammals.

Paperbarks are a major habitat type in themselves, forming crowded stands in seasonally flooded places (see Plate 3), and these are favoured nesting sites for ibis, egrets and to a lesser extent cormorants and some other predatory birds. Ferns often form a tangled understory in the shade below, and provide shelter for anything from land yabbies to snails and frogs. In the north, river pandanus is another fringing species that grows seasonally flooded, and where it hasn't been damaged by buffalo spreads into thickets that attract a wide variety of terrestrial animals as well as wetland species.

Redgum forests are less obviously a wetland habitat except when flooded, but this is the most widely distributed eucalypt and it is the single dominant species along countless thousands of kilometres of river, stream and ephemeral swamp. Apart from the many types of terrestrial animal dependent on redgum habitat, the role of these trees from the nutrient source provided by their leaves to snags that shelter fishes are central to the adaptations and lives of many aquatic and also terrestrial animals across their vast range. The illusion that redgum forests are barren of other plant life is the result of centuries of overgrazing, as can be seen in the rich understory communities in less damaged areas such as parts of the Barmah Forest on the New South Wales–Victoria border.

In clearer fresh waters pondweeds and eelgrasses form spreading underwater stands which are excellent shelter for smaller fishes and tadpoles, as well as being readily eaten by all herbivorous birds. In saltmarshes and coastal lakes (see Plates 6 and 8), seagrasses, watermats and widgeon grasses may be similarly important as food, and also as nesting materials for swans, as these plants often grow in shallow water so it does not take great effort to build up a nesting mound from them.

These major plant groups are often extremely abundant in places, but many other species would be similarly important if they were more common, whether as food, shelter or nesting material. Even at the level of the very small, examples of the myriad uses of diverse plants can be found; for example, the short sections of hollow reeds and rushes some caddisfly larvae adopt as their portable homes, while other types of insect larvae tunnel into the still-living stems and eat them from the inside, hidden from predators until they are ready to emerge as adults.

Underwater plants and water quality

One of the qualities popularly associated with submerged aquatic plants is their ability to 'oxygenate' water, yet a close examination of what they actually do tells a different story. Photosynthesis uses carbon dioxide and releases oxygen as a by-product, but the process only works by day. At night, plants reverse the cycle and actively use oxygen, and in warm, oxygen-poor environments with abundant plant life any animals that must breathe through their gills sometimes suffocate as a result.

In ideal situations such as the clear waters of Piccaninnie Ponds in south-eastern South Australia, the carbonate-rich water emerging from the springs below is low in oxygen initially, but is rapidly recharged by fine streams of oxygen bubbles produced by plants (see Plate 10). In the deeper areas these are mainly algae, but the clarity of the water allows even shallow-water plants to grow several metres below the surface, and on a sunny day the apparent visibility decreases because of the glittering shoals of tiny, ascending bubbles.

These are exceptional conditions, but most waters are much less clear so photosynthesis is sluggish at best, and on cold, overcast days may be negligible. Considered over a longer term, some plants may use up as much dissolved oxygen as they produce during their growing season; for example, many pondweeds die away during the hotter months of the year and the decay organisms that break them down use oxygen in the process.

In most wetlands there is an adequate amount of oxygen dissolved in the water even if there aren't any submerged plants present, and most of this comes from water circulation. Nearly all natural waters have currents, however slow, set up by even the lightest breeze, and as the water is in constant circulation most gas exchange is through the surface. The one time when this source of circulatory oxygen fails is on hot, sunny days in turbid waters, where an oxygen-poor heated layer riding on the surface prevents gas exchange.

Although plants aren't as important for oxygen production as generally believed, they do have other positive effects on water quality. The nutrient-stripping abilities of algae have already been mentioned, but most other wetland plants will do a comparable or sometimes even better job through luxury uptake, which means they can take up much greater quantities of some nutrients (including potassium and phosphorus) than for their immediate needs. Apart from allowing them to continue to grow when these nutrients become scarce, it also reduces algal blooms as algae need a similar range of nutrients, but don't seem to have the same stockpiling capacity.

Many wetland plants will also remove a range of less wholesome materials including some metals in their dissolved forms, as well as deal with everyday nitrogenous toxins produced as part of animal wastes. In the traditional version of the nitrogen cycle, it is bacteria that break down ammonia wastes from aquatic animals, changing this into similarly toxic nitrites, and then to much less toxic nitrates which can be used directly by water's edge and terrestrial plants. A number of freshwater plants, however, are now known to be able to take up and use ammonia directly.

Apart from their toxicity, these chemicals affect the ability of aquatic animals to take up oxygen through their gills, and a detailed account of these processes can be found in my earlier book *Sustainable Freshwater Aquaculture* (2007), as the nitrogen cycle is more critical in situations where animals are artificially crowded. In wetlands where an ecological balance has been achieved, most aquatic animal populations are low enough that none of these materials is likely to be present in threatening amounts.

The main exception is where there is a large gathering of birds, whether this is seasonal or more permanent, and the water below communal nesting areas and feeding grounds may contain such large quantities of droppings that very few aquatic animals can survive the resulting toxic brew. Although these nitrogenous wastes will gradually be broken down through the nitrogen cycle, the effect as they decay may last through the entire year from one breeding season to the next.

Plants as food

In wetlands where plants are a major component of habitat, a wide range of herbivorous animals will inevitably arrive to take advantage of them. Insects are probably the most prolific of these, whether it is their larvae or the adults which feed directly on the plants, though the huge populations of many smaller crustaceans feed largely on even more microscopic planktonic algae. Aquatic snails may also be abundant, even though there may only be one or two species of these present. In turn, dense populations of insects and snails will usually attract predators of their own, although this also depends on how densely the plants are growing.

I have seen a small lake so thickly overgrown with red water-milfoil that ducks and other waterbirds made no attempt to venture into the matted areas, congregating around a comparably sized lake several hundred metres away instead. As a result, the milfoil was seething with literally billions of amphipods (see Plate 11) which many of those birds would have been avidly feeding upon if they could only get to them. By the next summer, the milfoil was gone, the birds had returned, and amphipods were scarce once again.

Many other invertebrates also feed upon plants, and these are discussed in the next chapter. Of the native fishes, frogs, reptiles and mammals, few feed upon plants directly, and then only as an incidental item to their normal dietary preferences. Apart from invertebrates, the only other major group of animals to feed upon plants is herbivorous and omnivorous birds, and their activities can change the species composition of a plant-rich wetland.

There aren't all that many clear and unambiguous records of aquatic plants being used as food by waterbirds, despite an apparently extensive literature. For example, the two volumes of *The Food of Australian Birds* (1989) often don't specify which part of a plant has been seen to be eaten. In the case of rushes which are often mentioned, there must be a great difference in food value, texture and harvest method between the tiny, often hard-shelled seeds which are barely larger than some dust particles, and the tough, upright, wiry stems. Both seem implausible as food items, yet if it was recorded for certain that the seed was being eaten this would tell us a lot about the eyesight of the bird, and also might partly explain the prolific seed production of most rushes.

Other aquatic plant foods such as the various water ribbons (*Triglochin* species, see Plate 28) are only recorded vaguely as food for the wood duck, without specifying which part was eaten, and not for any other ducks, geese or swans, or even for the purple swamphen which forages on at least some of these species in preference to many other aquatic plants. Yet it is clear from my own observations and from conversations with many wetland managers that water ribbon leaves, seeds and tubers are readily eaten by many of these birds at least at some times, and may comprise the bulk of the diet at others.

It is not necessarily surprising that a major food item of this kind should largely go unmentioned in the literature, because it is soft and probably rapidly digested which makes it hard to identify, so it would rarely appear in stomach content assays. What is more surprising is that generations of written records of visual observation of waterbirds feeding should almost completely overlook this group of common and widespread plants, which is undoubtedly an important food source for many waterbirds.

In turn, this should act as a warning for when it comes to using published sources in planning or management, and not necessarily just for birds. For all of our apparent knowledge of many aspects of wetlands and their biota, there is far more that has not been published or noted, and even more still that doesn't turn up on internet searches! For this reason, it is worth making direct contact with specialists in the birds (or any other animals, for that matter) you have a particular interest in, many of whom have a wealth of personal knowledge which has yet to make it into print.

Plant zones and change

Most plants grow in distinct zones which can easily be seen around the perimeter of wetlands, following their preferred depth contours, whether these are areas that flood only occasionally, particular depths or tidal ranges, or deeper waters of a certain clarity. The seeds of some of these plants are also adapted to germination in a similar depth range, though those of many others may float and lodge at the water's edge before they sprout, with the plants creeping into deeper waters as they grow larger.

These zones aren't necessarily static, with the plants remaining trapped at one level and dying out if water retreats too far during periods of prolonged drought. Many aquatic plants spread by runners and will follow receding waters into areas which were previously deeper. Seed of other species lying dormant in waters too deep for germination will come to life once water levels drop enough, and colonising species with wind-blown seed may become abundant across areas of newly exposed mud.

If a period of drought does not last long, many of the running species will not have travelled far in pursuit of the receding waters, so they will also be able to move back as rains refill the wetland. However, if drought continues year after year such plants will eventually end up in much deeper areas, and though they will drown if the wetland refills to its previous level at least some of their seed will wash out of the deeper areas where it has been accumulating, and will germinate in shallower waters to start the cycle again.

In more stable conditions, plant communities change more slowly as those species that are well entrenched have usually already fended off any attempts by other species to establish themselves. Even after a drought which kills off the established plants, the seed bank in the soil is likely to mostly be from the species present before, though newly exposed plant-free areas are always an opportunity for new colonisers to establish. Unfortunately, at the present time these are as likely to be introduced weeds as anything indigenous.

Plants in need of habitat

There are also many native plants which come and go in wetlands, and without some mechanism of change to create new niches that suit them, most of these will gradually disappear as they don't seem to be able to compete against the more common species. Some are probably just vagrants outside the core parts of their range, for example swamp lily which is primarily tropical to subtropical, but also appears sporadically much further south. The largest and healthiest-looking plants of this species in cooler climates are found in churned-up farm dams, where high nutrient levels from stock droppings combined with a shallow, sun-heated layer of water around the fringes recreate its favoured habitats.

At another extreme, nearly all Victorian populations of the native form of water shield (see Plate 32) are found near the Goulburn Reservoir, where they form dense carpets of floating leaves over hundreds of square metres. It is likely that without the reservoir they would have died out, as most other southern populations had already done before European settlement. Although water-shield rarely sets seed this far south, the relatively stable water levels and temperatures of the banked-up backwaters in the area allow it to proliferate from rhizomes.

A major category of wetland plants that relies on change for survival is the annuals, most of which need relatively open sites clear of competition to establish and set seed, so they are most abundant in seasonally flooded areas or on regularly disturbed soils where longer-lived plants don't thrive. For other plants, too *much* change in the form of grazing and draining of wetlands may be the reason for their increasing scarcity, and some such as frogbit and maundia may be fading away largely unnoticed as a result.

It would be useful to learn more about the needs of such increasingly endangered plants and how to encourage them not only for their own sake, but also because we know so little about the animals that may be associated with them. The majority of these will be invertebrates, and as will become obvious in the next chapter, we know almost nothing about why some of these smaller animals are rarely seen, or at least recorded. It is not implausible that in many cases this is because the plant components of their preferred habitats are rare, too.

11
Invertebrates

In the zoological sense, the word 'animal' includes not just mammals but also birds, reptiles, amphibians, fishes, and even the bizarre and disparate array of mostly small to minute creatures called invertebrates. In this and the following chapters I have tried to emphasise what is special or distinctive about the various groups of animals found in and near inland waters, using examples from across Australia, and to bring out what is different about our indigenous biota to that found in other countries – as well as some important similarities.

This chapter looks at the myriad roles of the invertebrates, literally meaning animals that don't have backbones, though this oversimplified term covers by far the greatest proportion of all types of animals living. Essentially, these are all of the numerous and diverse creatures that don't have a bony, internal skeleton so they can't grow past a certain size, though some species which live entirely in water (which supports their weight) can become impressively large.

Although the emphasis in this chapter is on the diverse roles of invertebrates as food, as predators, and even sometimes as parasites, keep in mind that many of the less common species are worth conserving in their own right. This is not necessarily difficult as some of the examples that follow will show, and you may already be maintaining viable populations of some truly exotic creatures in a boggy corner of a farm paddock, or a shallow dam that dries out too often for your personal liking.

This is because many invertebrates have little or no ability to evade predators, as their major adaptations are to survival in habitats such as ephemeral pools where predators are rare, or arrive so late that their effects on population size aren't great. It is also true of many larger animals as will be seen in the following chapters, whose preferred habitats include the apparently empty ecological spaces where their predators and competitors aren't usually found. Some invertebrates will also breed up in huge numbers as quickly as possible, invariably much faster than any of their predators can keep up with, while others are cryptic – good at hiding.

Learning about smaller invertebrates can be done with very simple and inexpensive tools – a good identification guide, and a quality hand lens that will magnify around 10 to 12 times. If you become as intrigued by the bizarre creatures you find as I have been for half a century now, the next step is a stereo microscope that magnifies around 20 times, and this will also show you these fascinating creatures moving and feeding in three dimensions – more bizarre than even the aliens in most computer games!

There are so many types of invertebrates and so little is known about so many of them even now that readers will have to do much of their own identification and research if anything particularly interesting turns up. A good starting point to finding out who is studying the animals you are interested in is via invertebrate curators and specialists at the many museums around Australia, or via the internet as described in an appendix to this book, but don't expect other people to do any serious research for you unless you've turned up something they also find really exciting.

This is why the emphasis in this chapter is largely on the more common and ecologically important groups (particularly insects and crustaceans) about which more is generally known, though it should be kept in mind that even these may include hundreds of species in many very distantly related families, and that many others will be uncommon and poorly studied, or even rare.

Breathing and flight

The diversity of invertebrates is far greater than anything seen among the vertebrates, even though these larger creatures fill out the next three chapters, and not surprisingly the way they breathe varies as well. Many invertebrate species are true aquatics, so they collect oxygen through finely divided gills just as their ancestors have done for hundreds of millions of years, but others (including most insects) have had to evolve different abilities so they can leave the water.

The ways in which oxygen is taken up varies even within what look like natural groups, so among snails we find some that still breathe through gills, while others are more closely related to land snails and have lungs and will drown if kept submerged for too long. The insects have also had abundant time to develop a variety of strategies, from primitive groups such as dragonflies (see Plate 25) and damselflies which have always been associated with water and have aquatic larvae with gills, to the many other insects which breathe air directly, sometimes through the mouthparts, sometimes through a snorkel which is usually at the end of the tail, or even through a bubble of air carried down with them under water.

The smaller an aquatic insect or larva is, the harder it is for it to break through the surface tension layer at the water's surface, though this also means it becomes easier for it to get enough oxygen through even the most rudimentary gills. As there is far less oxygen dissolved in water than the 20% by volume in air, gills are a limiting factor and in more stagnant waters the common insects will be surface breathers, rippling the surface of the water as they dash up and break through to replenish their supply.

Some smaller insects such as backswimmers (see Plate 30) carry a bubble of air down with them, (held on their underside which faces upwards in this case because of the way they swim) and giving them a distinctive silver sheen. Semi-aquatic spiders (see Plate 24) also do this when they dive under a leaf, with the bubble trapped by a film of fine bristles over the abdomen. This allows them to stay under for some time in ideal conditions as oxygen diffuses into the bubble from the surrounding water, and carbon dioxide out of it, but they still need to surface for a new supply every few minutes, particularly in low oxygen conditions.

Breathing is also linked to flight – insects can fly because they can breathe air, but crustaceans could never develop wings because they mostly breathe through gills. Crustaceans are more comparable to fishes, which are also restricted to particular rivers or wetland systems in many cases, although many of the smaller species have also developed tiny, hard-shelled eggs that can be blown considerable distances in dust storms or survive passage through a bird's gut, and some of the larger species such as crayfishes have a limited ability to walk between wetlands.

Moulting and metamorphosis

The world swarms with arthropods – nearly all of them insects or crustaceans – in untold trillions, and even this may be an understatement. In both aquatic and terrestrial habitats they are the dominant animals in terms of numbers, biomass, and ecological impact as food animals, pests, parasites and competitors, yet the word is unfamiliar because we usually talk about their major groups instead – insects, crustaceans, spiders, scorpions, centipedes and a host of lesser known and even less popular animals. In aquatic environments the most significant groups by far are the insects and the crustaceans, sibling groups which separated hundreds of millions of years ago, but still have many things in common.

All arthropods are more or less bilaterally symmetrical, with one side of the body the mirror image of the other, and eyes and mouthparts at the front end of the body – this is also true for humans. The arthropod body, however, is made up of segments although this isn't always obvious where these have fused to produce more complex structures, and their 'skeleton' is more or less an external shell, jointed along the legs and feeding parts to allow movement. As the external shell isn't elastic, it must be split and shed whenever the occupant fills it too thoroughly, presumably feeling like a too-tight set of clothing at that stage.

Once the old shell is shucked off, the shell of the newly emerged animal is still relatively soft and is inflated to a larger size while it remains flexible, making space within for the softer parts of the animal to continue to grow until the next moult is needed. While the inflated shell hardens, an arthropod is particularly vulnerable to predators and will usually hide for this time, but it is also an opportunity for larger and harder-shelled species such as crayfishes and crabs to mate (see Plate 1). The male in these species may court the female for some time before moulting, and then guard her through the soft-shelled phase.

Marine crabs produce planktonic young that change through a series of truly surreal body shapes as they drift, eventually turning into small crabs that drop down to the bottom. By contrast, young freshwater crabs develop into miniatures of the parents before they are released – a reflection of the extremes of environment to which this widespread native is well adapted.

Insects are the dominant life form on this planet, filling most terrestrial niches suited to smaller animals that can't grow past a certain size because of the limitations of their breathing system. They are also among the most common freshwater animals as well, and many of them pass their larval (juvenile) stages under water. As insects moult and grow they eventually become adults via one of two very different pathways. Some change gradually from one moult to the next – these don't look all that different in shape from the adult form even in the beginning, and there is usually no great surprise after the final moult that leaves them as winged adults.

Dragonflies make the most dramatic change of this kind when they take to the air; one fine morning the nymph (see Plate 25) crawls up a stem of any plant that emerges from the water and escapes from its old skin, but then it spends a short time inflating the wrinkled-looking sacs on its back into full-size dragonfly wings. These harden quickly in the morning sunshine, and once they are at full strength and the dragonfly has warmed up it takes off on its first flight and never looks back.

Other insects such as mosquito wrigglers (see Plate 30) make a much more radical transformation called metamorphosis. The elongated, caterpillar-like larva is designed to be all efficiency underwater, feeding on tiny algae, bacteria and other small stuff, diving to the bottom of the pool to escape predators if disturbed, but spending most of its time near the surface breathing through a hollow tube at its rear end. When the time comes to become an adult, the larva takes on a very different, blocky shape, and using that as a kind of template it basically dissolves inside and reassembles itself into a very different guise – the flying mosquito, designed primarily to mate and breed.

It is interesting to note that some plants including a few wetland ferns have had such a long and contentious relationship with herbivorous insects that they use the moulting process against them, producing phytoecdysones that mimic moulting hormones, so that a larva feeding on their foliage basically moults itself to death.

Plankton

Plankton is an ecological category and not a clearly defined group of animals, a mix of the mostly microscopic, free-floating life found drifting in fresh, marine and even hypersaline waters worldwide. In both fresh water and saline wetlands, the bulk of the animal plankton (zooplankton) is usually made up of crustaceans, particularly copepods (see Plate 21), water fleas and their relatives (see Plates 11 and 21), and seed shrimps (see Plates 6 and 21), so the general ecology and breeding cycles of inland plankton are considered here before discussing the different groups of smaller crustaceans which are such a significant part of it.

Most planktonic crustaceans feed upon phytoplankton, bacteria and other micro-organisms, sometimes plant detritus, but particularly upon various single celled plants including algae and diatoms. There may be so many of these present that they colour the water, like the red saltpan algae in the strikingly coloured lagoons around the mouth of the Murray River, and as many smaller crustaceans are transparent, they may be coloured by the algae they feed upon.

Phytoplankton will usually move in the course of the day, drifting downwards away from the hottest sunlight and rising again towards the fading light in the evening, with the herds of planktonic browsers following. Zooplankton also respond to the wind, avoiding the roughest and most open waters by swimming downwards, or clustering in huge numbers along a lee shore so that there may be a band several metres wide in these more sheltered waters, and almost no plankton anywhere further out.

The greatest diversity and often the largest populations of plankton are in ephemeral wetlands, where the drying out process has killed off larger predatory animals such as fishes (see Plates 14, 20, 21 and 28). In cooler climates, if a wetland refills with autumn rains before the weather gets too cold there may still be time for breeding dragonflies and other predatory insects to fly in and lay their eggs, so their young can take advantage of the planktonic harvest soon to come, but the colder the weather when an ephemeral wetland refills the fewer insect predators will be around to take advantage of it.

Water beetles may be among the last insects in the season able to lay their eggs in cooler weather, so their predatory larvae (see Plate 28) can take advantage of the harvest, because they can also survive for some time in the soil of not-too-dry wetlands. However, other smaller predators will appear to take advantage of the glut of crustaceans including hydras, small and slender relatives of jellyfish and sea anemones with stinging, seizing tentacles, which may cover every surface they can attach themselves to like a furry, venomous mould. At another extreme, flying predators such as spoonbills may arrive as well, though like most birds they are too large to be able to detect let alone catch the smallest crustaceans.

A typical sequence in a freshly flooded wetland would start with the decay of any organic matter that has built up during the dry spell. Grasses and other fast growing terrestrial plants that have established themselves during the dry period will drown within a few weeks as the water level rises, decaying and providing fertiliser and food for blooms of bacteria and other decay-feeding organisms. These may become so dense that they form a smoky haze in the water, and it is only under strong magnification that it can be resolved into shoals of living things, which may be present in their billions even in a relatively small rainwater pool at the fringes of a sports field.

In turn, these will provide food for water fleas, copepods and seed shrimps, which usually hatch out from drought-resistant eggs and cysts within a few weeks. These animals live short lives, often only several weeks, but they make up for it with prolific breeding. In the case of water fleas most of these animals are females, producing still more genetically identical young for generation after generation. It is only as the wetland begins to dry out that this changes; salinity may begin to creep up as water evaporates, but even in the freshest pools organic wastes become more concentrated, predators appear, and less food is available. At this stage the water fleas begin to produce resting eggs designed to survive the coming drought, many of which will hatch into males at the beginning of the next flooded cycle.

Although the greatest glut of planktonic crustaceans is found in recently refilled wetlands, most larger and more permanent water bodies such as lakes and even dams also have a more permanent population of plankton, again including some species of water fleas but more often copepods. Seed shrimps may be more abundant near the shoreline and some species form a seething mist at the surface of deeper waters, either because fishes don't like their relatively hard shells or because they may actually be distasteful.

There are usually some water fleas or copepods in any slow-moving water body, even backwaters attached to streams, but where larger predators such as fishes can find them, their numbers are greatly reduced. It is worth noting that although native fishes are primarily carnivores, when they feed upon plankton they are also absorbing nutrients directly from the algae-filled gut of their prey, and this may be important for their health and successful reproduction.

The value of these 'hidden foods' has long been known to fish breeders, as fishes are often reluctant to eat captive-bred planktonic organisms such as water fleas, copepods and brine shrimp that have been fed on a limited and unnatural diet such as baking yeast. When given no choice they will eat them, but may not breed, or may produce eggs of poor viability as a result. By contrast, the same planktonic species raised on a diet of green water from a duck pond will be avidly eaten, and seem to act as a stimulus to breeding as well as improving egg hatchability and the vigour of hatchlings.

Smaller crustaceans

Identification to species level is difficult for many groups of smaller crustaceans, partly because they need to be magnified considerably to bring up the essential details for identification, but also because not all species have necessarily been described yet. The keelbacked water fleas (see Plate 11) are a good example, a distinctive group with a high, arched sail along the back. Treated in the past as a single species, closer study suggests they are a species complex which could do with further work, but the problem is complicated by other changes particularly in the shape of the head region, which varies even in a single species over the breeding season.

In lakes and other larger bodies of relatively still water copepods (see Plate 21) may be present in enormous numbers for much of the year, and other species of this group will appear in smaller and more ephemeral pools. Despite their great abundance, the wide range of inland waters they occupy, and their significance as food for a range of other animals, **copepods** don't have a common name. Some such as anchor worms are parasitic, with the female essentially a sort of living barb stuck in the sides of fishes, trailing streamers of eggs behind her.

The free-living species resemble a tiny but simplified shrimp with a distinct head and segmented body, on which clusters of eggs can be seen when females reach breeding size. There are three main groups. **Calanoid copepods** have long antennae and a relatively long, multi-segmented body, and are usually found in more open waters where a mix of up to several species feeding on slightly differently sized particles and algae make up virtually all of the plankton. **Harpacticoid copepods** have stubby antennae on an elongated and multi-segmented body, and are more often found on and around vegetation and sediment than free-swimming.

Cyclopoid copepods are the common species around the fringes of lakes, also with long antennae but a much shorter and rapidly tapering body, and these are mostly predatory species that feed on smaller planktonic animals though some will also feed on algae. Unlike the previous two groups, female cyclopoids carry a balanced pair of egg bundles rather than a single bundle.

Water fleas (cladocerans) are much more rounded in shape (see Plate 11), and carry their eggs within the (usually) transparent shell from which their swimming limbs protrude. Their resemblance to a flea is mainly in the jerky swimming action, sinking slightly then moving jerkily forward and up with each stroke. They feed on a range of single-celled algae and bacteria, and some will also browse on vegetable detritus presumably for its bacterial content as they are not likely to be able to digest it thoroughly.

Their resting eggs survive drought and will often pass through the digestive system of birds without harm, which is why they can be found in almost any water body where birds visit, but they are also light enough to be blown by the wind. These eggs are also long lasting, and have been hatched from the dry mud of ponds which have not been filled for centuries. Most other crustaceans from ephemeral wetlands also have tough and drought-tolerant eggs which can travel with winds or with birds, though less is known about how long they can remain viable in most cases.

Seed shrimps (ostracods, see Plate 21) look something like a seed, and something like a tiny swimming clam. They can even be found in the most ephemeral shallow pools which dry out within weeks, because the two halves of the shell in many species can close together tightly enough to protect the living animal from drying out. This gives seed shrimps a head-start over other smaller crustaceans that hatch from eggs, and must take at least several weeks to reach a breeding age, so a viable population of the most drought-tolerant ostacods can start breeding within days of a grassy pool filling. Under strong magnification the shells of many seed shrimps are distinctively sculpted or pitted, and these patterns can be used for identification.

Clam shrimps (conchostracans) are generally much larger than seed shrimps though much less common, some reaching more than a centimetre in length. Their shells are often sculpted with growth lines that make them even more clam-like in appearance, but the animal within is very different from a clam, with a segmented body and many pairs of swimming and filtering legs. Like seed shrimps they feed on a variety of algae and possibly also detritus.

Scuds and sideswimmers (amphipods, see Plate 11) have flattened sides and distinctive arched backs. There is a diverse range of native species found in mostly darkish and protected habitats including among decaying plants and in crayfish burrows, feeding mainly on vegetable matter though they will also scavenge dead animals. Some species have left the water and live among moist decaying vegetation on land – these are the so-called leafhoppers. Although relatively hard-shelled and not always easy to find, amphipods are most abundant where fishes are absent, and prefer more permanently wet places that don't dry out completely as they can survive partial drought by burrowing into organic matter.

Water slaters (isopods) look superficially similar to scuds, but are flattened from top to bottom, like their terrestrial relatives the garden slaters. Their feeding habits are also similar though not all of them will scavenge, and some are more likely to be found in semi-saline environments, while others are semi-aquatic and associated with backwaters that dry out at times.

Shield shrimps (notostracans or tadpole shrimps, see Plate 20) could be counted among the larger crustaceans as they reach lengths of several centimetres, but their relatively short lives and adaptation to ephemeral wetlands (including drought resistant eggs) place them squarely with plankton. The front half of the body is protected by a soft shield, and they swim rapidly if rather erratically with the many pairs of legs along their bodies. There are two species, one of which appears from winter into spring, the other in habitats that fill in autumn.

Fairy shrimps (anostracans, see Plate 20) reach a comparable size with similarly drought-tolerant eggs and the two are often found together in areas such as shallow dams and ditches grazed by livestock. Apart from keeping other plant growth down, cattle in particular provide large quantities of partly decayed organic matter in the form of manure which may be a direct source of food, though with their rapid growth rate it is likely both shield and fairy shrimps also feed on the even more readily assimilable bacteria associated with the wastes.

Brine shrimps are closely related to fairy shrimps, but are much smaller and are only found in salt lakes and other relatively still, saline waters. There are several species of native brine shrimps (see Plate 6), but the northern hemisphere species has also been introduced and is common in some places. Like fairy shrimps these retain a basic and primitive crustacean design, an elongated body with many segments, and not too much specialisation or variation between the many pairs of legs.

Larger crustaceans

Many aquatic crustaceans can reach an impressive size, particularly shrimps, prawns and crabs, and with over a hundred species of freshwater crayfish, our indigenous species compare in diversity to the even larger range in the far better-watered Americas. These include both the largest and smallest freshwater crayfish in the world, specialised for life in diverse conditions from burrows deep into the watertable, to cold mountain streams and muddy, sluggish waters in lowland areas. Although the bulk of their diet is usually vegetable matter, they will also scavenge meat and may actively hunt and kill sluggish fishes at night, and where they are abundant may determine what other species can survive and breed in the same waters. In turn, they are also an important and favoured food for many larger fishes, predatory birds and even humans from aboriginal peoples to the present day.

Spiny crayfish (see Plate 11) favour cool conditions and reasonably clear waters, and as they are slow growing and many of them reach a large size, overfishing has depleted many populations. The Murray River spiny crayfish is second only in size to the Tasmanian spiny crayfish, which is known to reach 4 kg and 60 cm, though much larger individuals have been reported in the past. Each neighbouring set of river systems in eastern and south-eastern Australia has its own distinctive species, often spectacularly coloured in reds, blues and vivid greens. These now-separate rivers were still part of larger catchments during the last Ice Age, but the places where they ran together are now mostly under the risen sea.

Smooth crayfish are lowland animals, tolerating a wide range of temperatures and salinities, many of them able to dig deep burrows as a refuge during periods of drought. Yabbies, marron (a rough-shelled 'smooth' crayfish) and redclaw are important aquaculture species, and all have the potential to invade a wide range of new river systems; yabbies and marron have already escaped from farms on Kangaroo Island in South Australia, and are found in such large numbers there that there can be no doubt they have changed the underwater ecology irrevocably.

Marsh yabbies (see Plate 9) with their deep bodies are often confused with land yabbies, but their tails aren't much smaller in proportion than those of smooth crayfish so they can swim well, reflecting a more aquatic life cycle. Although these are usually found in shallow, ephemeral wetlands (which may become quite saline as they dry out) and burrow readily in adverse conditions, they breed and feed most actively during the wetter months when their wetlands and streams are full.

Land yabbies are an extreme adaptation and are rarely seen as they only leave their burrows by night, and not often even then. These are primarily scavengers and foragers with shrunken tails, which combined with their exaggerated claw size gives them a very strange appearance if you are only familiar with aquatic crayfish. Often found in relatively high areas, their mud chimneys are a distinctive feature around springs and in open ferny gullies, dropping steeply down into the watertable which may be far below in summer.

The **freshwater crab** (see Plate 1) is the only widespread freshwater species of this group in Australia, also known as brownback, arid-zone and desert crab though it is also found in coastal streams and swamps in much wetter coastal areas. Well adapted to extreme conditions in inland areas, brownbacks will seal themselves into a deep burrow during times of drought, and unlike marine crabs with their complex set of surreal-looking planktonic larval stages, the young develop into miniatures of the adults protected under their mother's tail until they are ready to forage for themselves.

The **estuarine crab** is a smaller species of false spider-crab, a group usually found in estuaries or the sea, but enters much fresher waters so it may be found a long way inland, even in slightly saline lakes and streams, where it is extremely hard to see among snags. It is easily recognised by its flat, button-shaped top, and the spike-tipped legs are designed for climbing.

Diverse species of **freshwater shrimps** are found across the entire mainland, but the most widespread and abundant is the glass shrimp (see Plate 11) of eastern Australia and Tasmania. An adaptable animal, it is found in slower moving streams, lakes and backwaters, even in ephemeral streams that dry out to just a few waterholes where the shrimp can breed, but like most other larger crustaceans they are very sensitive to a range of toxins including heavy metals so their presence is usually a sign of good water quality. Several blind shrimps are known from caves in northern Australia, including one that occurs with the blind gudgeons and cave eels of the North West Range (see Plate 2).

Freshwater prawns (or long-armed prawns) are not all that different in general appearance to shrimps, but are mostly much larger and the territorial males have long, slender claws which rival the body in length. The more tropical cherabin which is also found in South-East Asia has been commercially farmed, though its various larval stages have specific feeding needs which take some skill to satisfy. The southern freshwater prawn is much more widespread in various forms, and extends through the Murray–Darling into Victoria. These are animals of slow-moving waters, usually found near cover including dense vegetation and masses of snags.

Insects

Insects are found in almost all habitats crustaceans hadn't already claimed long ago, and have had a free hand at colonising all suitable habitats for arthropods above the water's surface, where few crustaceans venture. This is where many adult insects live, hunt or mate, though many species from primarily terrestrial groups such as beetles and bugs have moved back into water, and the aquatic larvae of many insects live under water and breathe under water through gills. The more ecologically important aquatic insect groups are described below, but relatively minor players such as the few aquatic caterpillars and other minor players such as spongeflies and toebiters are not considered here.

Most insects can fly when adult, which is why they can be found even in tiny pools a long way from larger and more permanent wetlands, and some take advantage of such isolated pools as predator-free breeding grounds. This includes the invertebrates of most concern to many people living near wetlands, **mosquitoes**, which apart from their irritation value also happen to be vectors for some of the worst diseases affecting human beings and many other animals. The worst of these diseases aren't in Australia as yet, though with our close proximity to South-East Asia and international travel at an all-time high it would be foolish to become complaisant about them.

Many urban wetlands are routinely sprayed by ignorant councils to destroy what is imagined to be the source of mosquitoes, but these usually breed in small pools of water. Only the female bites as she needs a rich source of nutrient in the form of blood to produce her floating rafts of eggs, and although she may not have much of a brain, she instinctively knows that the larger a body of water the more likely it will be to have predators living in it. Small fishes of all types will eat every mosquito wriggler (see Plate 30) they can find, as long as they have reasonable access to all of the smaller pools in which a mosquito is most likely to lay eggs.

This means there is no point in spraying around the fringes of larger water bodies, because there won't be any larvae there. In urban areas the most likely breeding sites are created in suburban yards: a half-full bucket of rainwater under a deck, old water tanks, even the small pools formed in nests of old flower pots and tyres. With no predators present, it is possible for a single such container to breed thousands of mosquitoes each year. Even a garden pond with no fishes may be a breeding ground, though once wrigglers are present in a more permanent water body of this size, predacious insects such as backswimmers will often fly in to fix the problem.

Midges are related to mosquitoes, but not all of these are blood-sucking pests, and **chironomid larvae** (see Plate 31) are often abundant in oxygen-poor environments. Their bright red or orange blood makes them among the most conspicuous insect larvae, especially as they often move around in water by lashing and coiling frenetically. Most of these larvae feed on algae and bacteria associated with detritus, and as they are readily eaten by most fishes they are usually only found in fairly oxygen-poor environments, and isolated backwaters.

Other fly larvae associated with water are much less visible, particularly those of **phantom midges** which are almost completely transparent, and though they may be present in their hundreds drifting near the water's surface, can only be seen at certain angles of light. Diverse other flies from many distantly related families also have aquatic larvae, though the adults may be fully terrestrial and only return to fresh waters to mate and lay eggs.

Dragonflies and damselflies are the most widely recognised insects associated with water, noted for the beauty of their flight and spectacular manoeuvering which rivals that of a helicopter, but uses far less fuel. Their basic design hasn't needed any improvements in the 300 million years since they took to the air, and both adults and larvae are among the top predators in and around more permanent waters. Dragonflies (see Plate 25) are generally more solidly built and often much larger than damselflies, resting with their wings outspread, while most damselflies fold theirs tidily together along their back.

Young dragonflies are called mudeyes in Australia and are an ambush predator, with peculiar toothed jaws that fold away tidily under the head, but can be lunged out to a surprising distance to seize their prey. Their swimming style is equally interesting, as they breathe through gills inside the rear end of the body, and can also blast water out of the gill compartment to shoot forward in jet-propelled jumps.

Mayflies are almost as ancient a group, and their larvae are among the most abundant insects in some lakes and many streams, where they hide in large numbers among rocks, feeding on algae and plant detritus. The nymphs of many species transform into short-lived adults within a matter of hours (or sometimes days) of each other, and can take to the sky in such huge numbers on their mating flights that they look like billowing smoke. These provide an abundant but short-lived source of food for almost any smaller aquatic carnivore, particularly some fishes which may glut on them to the exclusion of all other foods for a few days.

The aquatic larvae of **stoneflies** are superficially similar to mayfly larvae, but with two-pronged tails rather than three-pronged. Some species are carnivorous rather than herbivores, but they are otherwise found in similar stream habitats, and both larvae and adults are longer-lived, so the adults don't necessarily emerge at the one time.

Caddisflies also have aquatic larvae though most of these are cryptic and are rarely noticed except by those who know where to look, as they are mostly disguised in a cocoon or case. These are constructed in many different ways depending on species, the simplest being just hollow sections of reed the animal can retreat into when disturbed, but other more elaborate cases are made of silk and may be covered in stones, various types of vegetation, and even formed into a shape like a snail shell. Particularly common in streams, caddisfly larvae are sensitive to a range of environmental problems, so their presence is usually regarded as a sign of good water conditions.

Beetles occupy a diverse range of niches in and around water, and most aquatic species also have aquatic larvae. Many of the larger species are active predators both as adults, and also in their larval stage which looks a little like a caterpillar with massive piercing jaws (see Plate 28). The larvae don't just feed on other invertebrates, but can even catch quite large fishes, and smaller tadpoles are a regular part of their diet, but there are also some aquatic beetles that feed on vegetation and detritus.

Whirligig beetles can dive but spend most of their time on the water's surface in groups, idling around each other as they wait for diverse food items including smaller animals (whether dead or alive) to fall onto the surface, at which stage their activity becomes frantic as they compete to locate the potential prey item first. Not all wetland beetles are so obviously streamlined for aquatic life, and many blunter-bodied species also forage around the water's edge and in overhanging vegetation.

The lifestyles of aquatic **bugs** are just as diverse as those of aquatic beetles, and like dragonflies they go through a gradual series of larval stages that look increasingly like the adults, rather than change abruptly through a metamorphosis. Only a few of the more conspicuous aquatic bugs are described briefly here.

Among the most widespread groups in relatively still waters are the **backswimmers** (see Plate 30), with oar-like limbs used to maintain their upside-down position below the surface. These are easily recognised by their silvery, cigar-shaped bodies from above, and bright red or pink eyes from closer up. The wingless larval stages don't look all that different from the adults though they are often a paler colour, and all stages are active carnivores feeding on small invertebrates of all kinds. They should be handled carefully as they have a sharp sting comparable to that of a bee in intensity, but fading faster.

Water boatmen (see Plate 30) are superficially similar to backswimmers but swim with their green, camouflaged backs upward and feed on vegetation as well as any smaller invertebrates they can catch, though they seem much less efficient as predators. While many fishes avoid eating backswimmers presumably because of their bite, water boatmen are only abundant in waters where fishes are uncommon or not feeding actively.

Water needles (see Plate 30) and their broader-backed cousins the so-called **water scorpions** are also bugs despite their common names, and are among the first colonisers of new farm dams, sometimes arriving as soon as there are enough insects to feed them. Despite their clumsy rowing motion through the water, they are skilled ambush predators and can even capture small fishes, seizing with the forearms and then stabbing with their pointed beaks. At times messy tangles of water needles can be found floating on the surface of a dam, looking like so much twiggy rubbish, and these are presumably some kind of mating aggregation.

Water striders are among the most fascinating and conspicuous aquatic insects, moving jerkily over the water's surface as they hunt for any smaller animals and detritus on the surface. The ripples they generate are also used for communication, and I like to imagine I can almost understand some of the diverse and varied messages they send to each other from the circles of light they create on moonlit nights, when they are particularly active.

Other arthropods

A few **spiders** are semi-aquatic, living around wetlands and streams and hunting on their fringes. These come from two different families, but are superficially similar. Aquatic wolf spiders are generally blunter and heavier-bodied, an aggressive animal the bite of which can cause serious tissue damage, while the more slender and elegant fishing spiders (see Plate 24) are longer-legged and can skate across the water surface in pursuit of prey, which they detect through ripples.

Both groups can dive under water but don't usually stay there long, and females carry their substantial egg masses with them as they hunt. Aquatic spiders will tackle surprisingly large prey including fishes, and I still remember my fright the first time I saw a big wolf spider of a species I had often picked out of nets by hand, seize a large fish in the same net and run up the side before jumping off with it.

Freshwater **mites** are mostly small and unlike their relatives such as ticks and other unpleasant creatures on land are mostly harmless and bumbling swimmers in slower waters and ephemeral pools. The young stages are parasitic on aquatic insects which is how they find their way to new waters, but the adults are predators on small planktonic animals. The species are difficult to tell apart without a microscope, and many are brightly coloured in red or orange presumably as a warning of inedibility to potential predators, as fishes will not even try to eat them.

Molluscs and worms

Molluscs aren't anywhere near as abundant in inland waters as in the sea, but water snails can be found in most inland waters where there are plants, and some even in saltmarshes where usually feed on anything they can find, from the bodies of tiny crustaceans to algal mats. Most aquatic snails are fairly small, and some of the smallest have greatly reduced shells as they are too small to attract predators, so they look like a tiny, very exposed limpet.

Although snails can be abundant in some environments and their gelatinous egg masses may cover the submerged surfaces of aquatic plants, they are mostly noted as carriers or intermediate hosts for some serious parasites such as liver fluke which affect humans as well as livestock, another good reason to fence cattle out of wetlands and dams, and provide them with clean water through a siphon or pump to a trough instead.

Freshwater mussels are usually only noticed as dead shells on dry creek beds or among the remains of water rat middens, which are one of their most successful predators, but may be among the most common larger invertebrates in streams and lagoons. On shallow sand bars in a moving stream there may be hundreds buried in every square metre just under the surface of the gravel, and they were once regarded as an important food by aboriginal peoples, though they are a little bland for contemporary tastes.

Although many mussels look superficially similar, they have very specific habitat requirements, basically falling into two ecological groups – those that only thrive in moving water, and those that prefer still conditions (see Plate 14). The slower-water species from inland rivers are exceptionally drought-tolerant, and can seal their shells tightly and survive (if protected from the hottest sun) for a year or two. Young mussel larvae attach themselves to the gills of fishes as a way of being transported to new locations, dropping off when they turn into a miniature of the adult. Smaller freshwater bivalves are usually called cockles, and may be even more abundant in the fine silt of lakes and stream beds.

Worms are only mentioned here in passing, as their identification is a highly skilled art, yet they are also important indicators of water quality. In polluted waters with an extremely low oxygen content they are among the very last survivors (see Plate 31), and in the absence of anything that feeds on them may form a living carpet. Their generally rich-red colouration is from the oxygen carrying pigment haemoglobin which also colours our blood red, and may not be as vivid in species from less extreme environments.

Worlds of the very small

Many of the most abundant and interesting invertebrates are much smaller than water fleas and copepods, which can at least be recognised with the naked eye, and some are only visible as a colour in the water, or as a cloud of what looks like dust. To even see the most basic details of these creatures you need a reasonable microscope, so most of them have only been considered in a general way earlier in this chapter – as part of the things plankton feeds on.

Rotifers are considered here as an example of just one group of these many unfamiliar creatures, as these are among the most abundant, characteristic and curious of the miniatures, and are also a startling illustration of how far removed the very small biota of wetlands can be from our conventional ideas of an animal. These wheel animalcules are among the small stuff that breeds and feeds in swarms among decaying vegetation in recently flooded places, in turn being fed on by smaller crustaceans and some aquatic insect larvae.

Rotifers are remarkably variable, so it is hard to define the features that make them a natural group, though they are usually said to have a head, a body (generally elongated), and often a foot though the planktonic species don't need one. Part of their fascination is that the individual animals are around the same size as many single-celled aquatic organisms, yet they are made up of hundreds of cells that form complex internal organs.

The mouth is in the head, and may be surrounded by various specialised feeding organs, and also a ring of cilia: hair-like structures which beat in sequence to propel the animal or draw food to the mouth, creating the swirling action that their common name of wheel animalcule suggests. Rotifers are mostly omnivorous (though some are carnivores), feeding on any debris that will fit into their mouths, and catching smaller life forms without bothering to decide whether these are animals or plants; others may even suck the fluids from algal cells.

Most rotifers are females, and males are unknown in some species; these are simpler creatures than the females as their only real job is to fertilise the eggs, though fertilisation isn't always essential. Females can hatch live young like themselves as long as conditions are suitable, but in adverse conditions produce a different kind of drought-resistant egg which must have a dormant period before it will hatch. This is such a successful formula for survival that rotifers are abundant in almost all types of freshwater wetlands, from the most ephemeral puddles to permanent lakes, and may even swarm in their thousands in a forgotten, half-empty bucket under your sundeck, where they will become an important food for some types of mosquito larvae.

12

Fishes

The freshwater fishes of Asia, Africa and the Americas have evolved into thousands of species over uncounted eons, yet almost none of the major groups such as carps and tetras have reached Australia, simply because they can't tolerate sea water. Of the catfishes, perhaps the most diverse group of freshwater fishes worldwide, only a handful of species derived from two marine families are found in our fresh waters, and even the very few marine members of the otherwise very successful and species-rich cichlid family haven't made it here.

Several indigenous species have evolved from true freshwater families, in the tropics the two saratogas and Queensland lungfish, and the enigmatic salamanderfish (see Plate 23) of south-western Australia. Nearly all other inland fishes, however, are descended from marine ancestors and many of them still have a marine stage in their life cycle, so they remain tied to the sea and to near-coastal habitats. In the south, the most familiar and species-rich family is the galaxiids (sometimes called native minnows), many of which breed in estuaries, and although these are also found on all other southern hemisphere continents they are most diverse in Tasmania.

The greatest overall variety of freshwater fishes is found in northern Australia, and most of these are descended from groups which have moved south from Asian shores, though in many cases they have had ample time to evolve into distinctive Australasian families such as rainbowfishes and the related blue-eyes. Others including the grunters have diversified in the inland waters of Australia and New Guinea, and although a few species are widespread elsewhere, Australasia is a major evolutionary centre for this family as well.

Regardless of their origins, like most true freshwater fishes worldwide nearly all native species need permanent water, and disappear even from major river systems if these dry out completely at times. In coastal areas depopulated waters will be recolonised by species which can survive sea water for at least a part of their lives, and the ability to do this on our drought-prone continent may be what has prevented so many of them from cutting their ties to the sea completely.

The freshwater lifestyle is perhaps most deeply established among many of the northern fishes, as the torrential rains and associated flooding of the wet season allow dispersal over great distances in a good year, without need of a marine stage. Many of these fishes, especially the smaller and younger ones end up trapped in waters that dry out once the floods are over, so the adults of some species breed prolifically to maximise the chances of some of their young finding their way to more permanent waters, while other less fecund species put a lot more care into raising a smaller number of offspring to a larger size.

It is important to keep in mind that Australian freshwater fishes are, almost without exception, primarily carnivorous. Some will eat vegetable matter or even certain fruits as part of their diet, at some seasons or stages of their lives, and while others may need a certain amount of algae in their diet this is usually sourced as a 'hidden food' from the gut contents of the invertebrates they feed upon. This is why omnivorous fishes such as silver perch, used extensively in aquaculture don't thrive on a diet too rich in plant material, despite their reputation as one of the few indigenous vegetarians among fishes.

Freshwater fishes needing access to the sea

Many estuarine fishes move freely between fresh waters and the sea, especially during their young or non-breeding stages, but their ability to tolerate non-marine conditions is little more than a luxury allowing them access to a greater range of feeding habitats. With time, some of these will increasingly move into fresh waters (as some of their relatives already have), but they need not be considered here as it probably wouldn't affect them if every river in Australia dried up permanently.

All freshwater fishes that need access to the sea for at least part of their life cycle breed either in the sea itself, or in tidal estuaries, and maintaining healthy populations of these is mainly dependent on their being able to move freely up- and downstream. Fishes of our larger, inland rivers also need to be able to move freely upstream, especially when the urge to spawn has been triggered by floods, and ways to counteract the effects of weirs, dams and various other river 'improvements' that block their passage have been discussed in Chapter 7.

Very few fishes have the ability to pass human-made obstructions, eels, climbing galaxias and some lampreys being among the few exceptions as their young can climb vertical rock faces. It is usually the upstream climbing habit that is considered most critical to the free passage of such species. However, even non-climbing species moving downstream can be physically damaged if they aren't adapted to the turbulent water conditions below weirs and dams – and few of them are.

Of the native species with some ability to climb, the complex life cycle and habits of the short-finned eel in particular were discussed in Chapter 1 (see Plate 16), but their interest extends far beyond the very different habitats they need for different stages of their lives, and the body shapes they tailor to exploit those habitats. Eels were also once a major predator, taking a wide range of foods from smaller fishes and crustaceans to young turtles, frogs and even small ducks.

Early accounts suggest they were a significant and sought-after food for indigenous people, but over the last few decades a combination of over-fishing and possibly competition from the introduced trout have diminished their numbers appreciably. Ironically, they are now probably more common in farm dams in some areas than in nearby rivers, having taken advantage of their rock-climbing ability to travel overland on wet nights to this new resource, rarely occupied by other large carnivores.

Although still moderately abundant in places, eels may be endangered by commercial fisheries in the long term, as no-one has the slightest idea whether they are being harvested sustainably. Consider that it takes around two decades or even longer for a large female eel to reach maturity, and the unknown odds of her survival on the journey to the Coral Sea. No-one knows how eels meet up in the depths of the sea, but unless they can find each other over great distances and aren't fussy about their mating partners or conditions, future generations could be suddenly and dramatically reduced by present levels of fishing.

Young eels that survive the return journey from the Coral Sea are also commercially harvested from estuaries as they begin their trek inland, and until fairly recently large numbers of them were moved to landlocked lakes from which they were unlikely to be able to find their way back to the sea, creating fisheries where none may have existed before. Many of these lakes are now declining rapidly or dry, and the eels in them are dead, victims of prolonged drought in many parts of southern Australia. It will still be decades before anyone can definitely show that other populations of eels are also in decline, let alone isolate the causes or take any remedial action, and by the time any action is taken it may be too late as has been the case for many other fishes and fisheries worldwide.

Most other fishes that commute between wetlands and the sea via rivers and estuaries are too small to attract commercial fisheries, and the main threats to these are river 'improvements' such as snag removal, and the ubiquitous barriers in their many shapes and forms. Like young eels, young climbing galaxias can also creep their way to above waterfalls and similar barriers, but most other galaxiids have a more typically fishy way of breeding and dispersing. Some of the purely freshwater species of this group are discussed later in this chapter, but most follow the older, sea-oriented patterns of the family.

The common jollytail was also mentioned in Chapter 1, one of a group of related species needing access to the sea. Although the adult fishes may spend most of their lives in fresh water, in spring at certain phases of the moon they move down to estuaries to spawn, laying their eggs on plants at the highest tide mark. Many of the spawning adults may die afterwards, while others will survive to breed in later years. Their eggs develop above the reach of aquatic predators, and will hatch with the next set of spring tides that reaches them around two weeks later.

The semi-transparent young move out to sea for the next few months of their lives, and may travel impressive distances. Common jollytail is one of the most widely distributed freshwater fishes because of its ability to cross oceans at this stage of its life, and the delicate-looking, transparent young have been caught hundreds of kilometres from the nearest land. When they return to estuaries and begin to migrate upstream they may be caught as whitebait, a minor fishery in Australia, but much more important and regulated in New Zealand.

The whitebait catch is usually a mix of several species of galaxiids, sometimes including the more-distantly related Tasmanian whitebait. Some of the estuary breeding species seem predisposed to adapt to a life in fresh waters, and if trapped in a lake or river so they can't get to the sea such landlocked species may breed along the fringes of streams during floods (see Plate 13). This may have been how several primarily freshwater species evolved, as has been shown in some landlocked New Zealand lakes, though spring flooding in most Australian rivers is much less reliable so there aren't as many places for permanent inland populations to develop here.

We don't know anywhere near as much about the breeding strategies of most other estuarine-breeding fishes, mainly because they are of no commercial importance and are correspondingly less studied. These include such oddities as the congolli, a rather flathead-like fish from a mainly marine circum-Antarctic family, which can be caught from the sea and dropped straight into fresh water without the slightest apparent effect – it may even keep on feeding as if it has not been disturbed. Although often found in streams this is a good example of a fish in transition with a mild fondness for, but possibly no real need of fresh water.

The cryptic lampreys are often assumed to be a type of eel because of their shape, but are jawless creatures only distantly related to true fishes. They are included here and in all fish identification references as a matter of convenience, as it is unlikely that a field guide to the three native lampreys alone would be a best-seller. Like eels, lampreys also need free access between marine and fresh waters and can also travel overland to some degree to get past rocky barriers but they spawn in pebbly streams, dying soon after, and their young feed by filtering algae and plankton from water and sediment until large enough to move out to sea. One little-known species remains in fresh water all its life, but the marine stage of the other two species is parasitic on sea fishes until mature.

Other families of fishes that include mostly true freshwater species, as well as the more tropical families that include species needing access to estuaries are discussed in the last section of this chapter.

Breeding strategies of freshwater fishes

We still know surprisingly little about the breeding habits of many native freshwater fishes, and some of their breeding habitats may be forever inaccessible to human beings, even if they live only a few metres out from the shoreline. What they all have in common is the need to maintain an adequate oxygen supply to the developing eggs, which usually includes keeping them away from silt. In many cases, successful breeding strategies require at least some parental care.

For example, the nurseryfish is an estuarine species in which the males carry grape-like bunches of eggs on a projection from their forehead, an adaptation that keeps them up in moving, aerated water and also well away from the sediments of the river floor. While a description like this gives the impression that we know something of its breeding and the way it lives, in reality we don't even know the full range of this species in northern Australia, let alone any other details of how it lives and breeds. Until some genetically modified biologist develops crocodile-proof skin, the ability to see more than a few centimetres through turbid water, and a desire to spend a lifetime exploring the muddy floor of every major river in northern Australia, none of this is likely to change.

On a continent of extremes, it is not surprising that many different fishes in a variety of climates are adapted to spawning after floods, and may require the stimulus of a flood for the final stages of egg and roe development to take place. Warm, summer floods are the signal for some inland fishes such as golden perch to breed, and their immediate instinct is to head upstream as far as possible before doing so: this makes it less likely that their buoyant, fast-developing eggs will be washed out to sea before they have hatched.

In the cooler streams of the south, the short-lived eastern little galaxias (see Plate 9) and its relatives in Western Australia are also triggered into breeding mode by floods, but during the cooler months. As the rains set in and floodwaters spill over surrounding floodplains, drowned terrestrial grasses decay to fertilise a bloom of micro-organisms suitable for the tiniest fry. By the time the fast-growing young need larger foods, water fleas and copepods will also have had time to grow and begin to breed, producing an abundant food supply.

The breeding habits of other fishes may be centred around what they attach their eggs to, and whether they need to look after or defend them once laid. Gudgeons and gobies attach their eggs to rocks and other firm surfaces, keeping them away from silt and its associated bacteria and fungi; shaded locations are often chosen as these aren't covered in a crop of algae. The male (or both the parents in some cases) not only defends the eggs until they hatch, but may also fan them to keep a flow of relatively oxygenated water up, and picks off infertile or infected eggs.

Rainbowfishes are more negligent parents, but breed prolifically over a much longer period of time to make up for their casual approach. On sunny mornings, the males attract their mates to overhanging or floating vegetation near the surface, where water temperatures are warmer, oxygen is more readily available, and development is faster. Threads on the eggs tangle into the vegetation, where they are suspended well above the silt layer and are also scattered so widely that infected or infertile eggs don't come in contact with those that remain viable.

Despite this being one of the most abundant and ecologically important groups of fish in warmer parts of Australia, and popular with aquarists worldwide, we still know very little about some aspects of their breeding biology. For example, the tiny fry are often very variable in growth rate, though no-one knows why, yet in many other types of aquatic animal that have very different growth rates between individual hatchlings there is often a reason once we know enough about the species to understand what it is.

This is why captive breeding of rainbowfishes by aquarists is not a useful conservation tool, as the fastest growing individuals are invariably selected and the slower ones discarded, even though we don't understand why this pattern is common to most (and perhaps even all) rainbowfishes. Further selection for the most colourful males leads to still more inbreeding, so most aquarium-bred strains of rainbowfishes that have been in captivity for 20 years and more look nothing like the wild fishes they are descended from, and invariably show increasingly poor fertility.

Mouth brooding is possibly the most extreme way to raise a smaller number of offspring to a larger size among native fishes, and is practised by three unrelated groups in Australia. Female saratogas and male salmon catfish carry particularly large eggs in their mouths, pumping oxygen-rich water over them day and night, while keeping them hidden from predators, and may also guard their large fry from predators once they have hatched. This is the ultimate in parental care, but has its cost as the incubating parent can't feed during this time, and is weakened by the experience.

Protection of fish habitats

Wherever possible, habitat protection should be the first priority in conservation, and although this is true for all living things in wetlands, it is even more so for fishes. This is partly because they are unlikely to spontaneously recolonise from elsewhere, unlike animals that are able to move overland or through the air. Even frogs have a better chance of recolonising a former haunt as long as there is another population within a few hundred metres, or perhaps even kilometres in some cases, but in the case of fishes the intervening land presents an insuperable barrier.

While many fishes (and also other wetland animals) are apparently well protected by paper laws, in practical terms these special privileges just make it more difficult to study them, yet rarely include any provision for protection of their all-essential habitats. Consider the Queensland lungfish, a species close to the rootstock of terrestrial vertebrates and which has effectively remained unchanged for 150 million years. Because of its evolutionary importance and relative ease of access to its habitats, the lungfish has been well studied compared to most other freshwater fishes, so it would seem an easy matter to protect it for all posterity. The breeding and feeding habitats are quite separate, and fish migrate moderate distances to particular breeding areas in slow-flowing waters, where they lay their large and sticky eggs among submerged plants.

Protected at State level for a century, the lungfish should have had every prospect of survival into the indefinite future. In practice, there seems to be a significant risk that the Queensland Government is in the process of slowly killing off the wild population, by damming the two small, coastal river systems where it is endemic. Already 80% of the known spawning grounds in the Burnett River have disappeared behind the Paradise Dam. As a token gesture towards ameliorating its effects, a special fishway was included so that lungfish can make their way up- or downstream, even though they are not enthusiastic travelers.

Despite a lack of evidence that it is being used, or is even suitable for lungfish use, a similar fishway is planned for the Mary River if that, too, is blocked by another dam. Until recently, the Government has been eagerly emphasising that there has been no 'apparent' change in the *adult* populations in the several years since the flooding of the Paradise Dam (that is, the adults hadn't dropped dead at that time), but there is also no sign that they are breeding.

More recently, these attempts to suggest that damming their river habitats is having no impact on lungfish populations was undermined by the deaths of unknown numbers of adult lungfish, swept over the North Pine Dam by winter flooding in 2009 – some were found wedged in trees!

Such events are likely to become increasingly common as global warming progresses, and pose a real threat to an iconic fish of international interest, subject to rigorous national and international laws enacted by the grace of Parliament for its protection. It makes you wonder what hope there is for small, uncharismatic fishes mainly used as bait, if they should happen to be become endangered.

When all else fails – introduction and reintroduction

Where habitat protection fails and a fish vanishes from one of its native haunts, it may be possible to reintroduce it – but only after whatever has caused its local extinction has been dealt with. This should be obvious, yet it is remarkable how few fish reintroduction programs fail to assess exactly what the cause of the problem is, or try to do anything about it ahead of reintroduction.

The problems besetting captive breeding programs for the lake form of the Eacham rainbowfish have already been discussed in Chapter 7. These included obviously futile attempts to return captive-bred populations to the lake, when introduced 'native' predatory fishes from outside the lake were most certainly the cause of its disappearance in the first place. No attempt was made to reduce predator numbers ahead of reintroduction, let alone remove them altogether, and not surprisingly the captive-bred fishes with their limited genetic resources, possible cargo of aquarium diseases, and even less selection for predator-wariness than the original populations had in the first place, disappeared without trace.

The only and extremely remote chance this particular population may have of survival in anything like its original form is to introduce it into a created wetland with similar conditions to the lake it came from, though whether this would be possible or economic to do is a moot point. Given the existence of vigorous wild populations of this species in rivers near the lake, and the dubious merits of re-establishment of a seriously inbred, possibly diseased aquarium population which is more likely to take shelter in a predator's mouth than run from it, there seems little incentive to make the effort to create a new and predator-free environment.

Equally ill-conceived was a recent attempt at reintroduction of the purple-spotted gudgeon (see Plate 12) by NSW Fisheries at Narrandera, using the offspring of a dozen or so fishes sourced from aquarists. With hundreds of unwanted offspring from these unprovenanced fishes (which may have included coastal forms from much further north) to dispose of, it was decided to release them into natural waters somewhere near Gundagai.

Unfortunately, this species has pretty much disappeared from almost all of its southern range long ago, the last wild fishes caught near the Hume Weir in the 1970s, and the odd fish that have reappeared in isolated places further south appear to have been swept down the Darling River from southern Queensland. The causes for their disappearance are not certain but are likely to include competition from plague minnows (which breed in similarly warm shallows), and altered water temperatures created by impoundments.

With no attempt to identify and correct for these factors, it is unlikely that a batch of random offspring of aquarium fishes will survive where wild populations have died out. If these also carry introduced diseases from the aquarium trade, such unplanned introductions may instead become a vector for new problems for other endangered wild fishes and frogs.

For other fish species and populations that are much closer genetically to the wild stocks they descend from, and threatened by processes that can't be corrected in the wild, a case could perhaps be made for introducing them into wetlands specially created for that purpose. This would depend on several other considerations, including whether introduction or reintroduction within a particular catchment is legal, and would certainly be subject to some kind of permit process as discussed in an earlier section of this book.

Release into natural wetlands where a particular fish has not been found in the past is not an option, because it will inevitably affect populations of other wetland animals, particularly some types of invertebrates. Many fishes will also prey on the tadpole stage of frogs, though to a lesser degree than is usually assumed (fish–frog interactions are discussed in the next chapter). As the Eacham rainbowfish example has already shown, introduction of some fishes can also drive other indigenous fishes in a natural wetland to extinction.

There is also little point introducing fishes to wetlands subject to flooding from upstream, if this is likely to bring in vermin such as plague minnows, guppies, carp and cichlids. If this is not an issue, the next major consideration must be whether enough is known of the biology to create a complete habitat for that species, able to sustain a sufficiently large and genetically varied population to avoid the problems seen in smaller, bottlenecked groups. In practice, this means a minimal population size of many hundreds, preferably approaching a minimum of a thousand.

The widespread misconception that water plus plants is all you need for adequate fish habitat has little or no basis except for the most adaptable fishes, few of which need any help from humans, and in the case of many species is just a slow way of killing them off. The most objectionable example of this is the not-infrequent cases where 'local' fishes are caught in a nearby creek and dumped into a dam, sometimes without even attempting an accurate identification of what has been caught.

Randomly caught mixes like this often include the highly undesirable plague minnow or its more tropical equivalents, and carelessness in checking the identification of each and every individual fish caught will introduce these as well. Equally objectionable is the assumption that every indigenous fish caught is suitable for introduction into enclosed wetlands from which it can't escape, yet in most parts of Australia at least some of these local species need access to estuaries or the sea to complete their life cycle, as has already been discussed.

Providing complete habitats

For many freshwater fishes, the ecological or biological aspects for successful establishment in a wetland may be far more subtle than this, and in many cases are still unknown. As an example, it took me more than two decades of working with the eastern little galaxias (see Plate 9), both in captivity and in the wild, to understand why populations of this diminutive fish often come and go with no apparent pattern. Introduced (or reintroduced) populations would almost always breed successfully, yet their young would vanish over the next few months, and usually there would be no sign of the species a year or two later.

Eventually, it became clear that neither young nor adult fishes seem to be able to recognise significant predators such as mudeyes, or had any instinct to avoid the microhabitats in which these are most likely to be common. In relatively undisturbed, natural regimes this didn't matter, because flooding was the cue for adults to move out to newly flooded areas where there were no such predators.

Most larger predatory native fishes are also relatively inactive at the cooler times of year and avoid shallow flood waters in any case, so the young could grow up in an environment that was effectively predator-free. Even as adults, many of the more reliably perennial populations of this species live out their lives in ephemeral wetlands where they may retreat to a few small pools where predators have little chance of survival if they follow; the flooded, underground parts of marsh yabby burrows are the most extreme of these pocket refuges.

Confirmation of predation as a major driving force in the disappearance of the little galaxias came from a few surviving populations in disturbed, urban habitats, where introduced competitors such as the plague minnow added further pressures to their continued survival. Here, the last remaining populations of this cold-tolerant fish are often found in shaded pools under bridges and road culverts, where poor light and low water temperatures discourage breeding dragonflies and the warmth-seeking minnow.

In the conditions under which the eastern little galaxias evolved, it survived by simply moving away from its enemies, breeding in shallow floodwaters during the colder time of year when most predators were mostly inactive. When the conditions it was adapted to changed so that its enemies could build up, or it was introduced into more permanent water bodies where there is no predator-free breeding area, the eastern little galaxias rarely lasted long.

Many other smaller species may have equally subtle or overlooked needs for their long-term survival, yet it is usually assumed that there isn't much about them worth studying. Academic researchers tend to concentrate on fishes that are regarded as endangered over much of their natural range, as these are most likely to attract funds for the detailed studies that will tell us what is going wrong. In practice, that means that a greater understanding of many native freshwater fishes is going to come from the work of those with a personal interest in them, rather than professionals who are more likely to follow where funding is available.

Major groups of freshwater fishes

This section is a brief outline of major groups of indigenous fishes which include at least a few species that complete their life cycle in an enclosed body of fresh or not-too-saline water. In other words, these groups include some animals whose complete habitat and breeding needs may *potentially* be met in a single place, always assuming that their biology is as uncomplicated as it is presently believed to be. Freshwater galaxiids and gudgeons are not included here as they have already been discussed earlier in this chapter.

Even though we know a moderate amount about these various species and groups, each summary which follows is little more than a thumbnail sketch, not including much detail of such things as precisely what they eat, their interactions and incompatibilities, or seasonal habits even if any of these things are known. Many groups including soles, flagtails, bullrout, swamp eels, longtoms, many grunters, freshwater garfishes, anchovies and herrings haven't been included because we know too little about most aspects of their lives to be able to cater for their possibly specialised needs in wetlands.

The **salamanderfish** (see Plate 23) of south-western Australia is a single, curious species of unknown origins, though it is believed to have been evolving in fresh waters for a very long time, and it has even been suggested that its closest relatives are the huge and predatory pikes of the northern hemisphere. A slender, scaled species with a flexible neck that makes it seem most un-fishlike, it was originally thought to be related to the scaleless galaxiids, but once its unique nature was recognised it became one of our most-studied inland fishes.

The salamanderfish is among the most drought-tolerant aquatic animals of any kind, as befits a species that has been evolving on this continent for perhaps 90 million years, burrowing into the sediments when the peaty and acidic ephemeral ponds that are its main habitat dry out. Water loss during this time is reduced by a slimy sheath that forms on the body, and the entire metabolism changes while it aestivates until the rains return. Although restricted to a narrow coastal belt this species is common in many places, including well-protected areas in national parks, but it is also a very adaptable beast and seems to readily colonise new places such as borrow pits along roads (see Plate 23) as long as these are in sandy, peaty soils, with suitably acid water.

The two **saratoga** species are relicts of an ancient group of true freshwater fishes, and are large and very distinctive predators in some tropical waters, particularly slower moving creeks and backwaters with overhanging vegetation. Both species can reach a metre in length and their large, upturned mouth accommodates a wide range of prey from frogs to other fishes, while the large eyes suggest their need for clear water though the more southerly species (also called spotted barramundi) is often found in more turbid conditions.

Usually solitary, spawning is triggered by rising temperatures just before the wet (around 30°C for the gulf saratoga, as low as 20°C for the southern species), and females brood the eggs in their mouths. They also protect the young after they have hatched, until these become fully independent at around 4 cm long. As they are excellent eating and also an active fighter when hooked, saratogas are highly regarded in sports fisheries and have been overfished in parts of their range.

Archerfish (see Plate 15) are also large-mouthed, surface feeding, tropical predators though they are much smaller than saratogas, and will also feed on floating fruits and flowerbuds. They are named for their ability to knock insects down from overhanging vegetation with a jet of water forced through the mouth, and larger fishes may 'shoot' accurately to a distance of three metres. The primitive archerfish is a freshwater species, but the other two are also frequently found in coastal waters, and one of them is known to breed in both fresh and brackish waters. Their floating eggs are small and produced in large numbers.

The **mouth almighty** is yet another large-mouthed predator, but its cryptic colouration allows it to lie in ambush among submerged plants rather than hunt actively. It is one of a relatively few freshwater species of cardinalfish, a large marine fish family and among the few marine groups that brood their eggs in their mouths. Mouth almighty is mildly infamous as one of the several predatory fishes introduced into Lake Eacham, resulting in the extinction of the lake form of Eacham rainbowfish.

Pygmy perches are similar to mouth almighty in profile, but are not closely related and are much smaller. The several species are found in slow moving and still, fairly clear and permanent waters of southern and eastern Australia. Although often abundant pygmy perches don't form schools, and may be a significant predator on aquatic insects, larvae and smaller crustaceans as well as an important food item for larger fishes.

The males develop more intense red and yellow colours and more strongly marked patterning in breeding season, which starts in winter for the south-western species, and in spring in the cooler waters of the south-east. Spawning is usually among plants or over relatively sediment free surfaces such as algae over mud. The eggs hatch in days and young fish grow rapidly, so they are often already at breeding age by the following spring.

Blackfish were once an important predator in the cooler, clear waters of south-eastern Australia but are now greatly reduced in numbers in many places, partly due to overfishing as they are very good to eat, but more seriously as a result of competition from introduced trout which favour similar habitats. Elongated, soft-skinned and nocturnal, they breed in hollows among rocks and sunken timber, so river 'improvements' such as de-snagging have also affected their abundance.

As they aren't prolific breeders it may take populations a very long time to recover from any setbacks, and although they will breed in dams with appropriate furnishings, there are legal restrictions on transferring fish from wild populations. The equally curious **nightfish** of south-western Australia is probably a fairly close relative though it is in a different family, and is also a nocturnal carnivore.

Freshwater basses and **cods** include the largest and most significant predators of inland Australia, with a few species also found in coastal catchments and rivers. Many of these are favoured targets for anglers, including the enormous Murray cod and golden perch. They are also among the most important species in freshwater aquaculture, though stocks from these sources come from genetically limited backgrounds.

Conservation efforts for some of these fishes that have become significantly depleted in the wild should focus on habitat and fish passage restoration, rather than hatchery production from a limited number of captive-bred animals. As almost none of these giants will breed in dams and other still waters which never flood, and they will eat almost any other animal which will fit into their mouths, they are not recommended for stocking in farm dams or created wetlands for any conservation purposes whatsoever.

Grunters are named after the awful sounds made by some species when lifted out of water, and include several species that are raised as aquaculture animals, of which the silver perch is perhaps the most-studied native freshwater fish. It is a flood-triggered spawner laying floating eggs that hatch quickly, though many other members of the family are known to lay eggs that sink into gravel. Although all species are basically carnivorous, some will feed on fruits and other vegetable matter when available. The widespread spangled perch (see Plate 1) has already been noted elsewhere for its flood-dispersal habits.

Glassfishes are generally small tropical fishes which are most active at night, and one species was once regularly found all the way down the Darling River to the warmer parts of Victoria and into South Australia. Although some are almost as transparent as their name suggests, others may have strongly marked patterns. The smaller species often school in clear, slow flowing or still waters in huge numbers, living and breeding among submerged plants, and only the rainbowfishes are likely to be more abundant in many tropical and subtropical habitats.

Rainbowfishes are a primarily tropical to subtropical family, often found in abundance and sometimes in mixed schools of up to several species. Because of their numbers in many of the warmer areas of this country, and the schooling habit of the younger fishes in warm, shallow waters, they are an important food for a wide range of carnivorous animals just as tadpoles are. Males (see Plate 15) display near the surface at the fringes of dense submerged vegetation where the eggs are laid, hatching within a few days, and there is no parental care; some females may even eat the eggs of others as they are being laid. To compensate for this, hundreds or even thousands of eggs will be laid by each female through the warmer months, and survival rates in ideal conditions can be high.

Usually found within a few hundred kilometres of the coast, there are also rainbowfishes in desert regions, while the Murray River rainbow has spread south to central Victoria via a tortuous route initially following the Darling River from southern Queensland. From the junction with the cooler Murray waters it has moved upstream until reaching the Goulburn River, then headed south into new and distinctly temperate climates, evolving as it went into a blunt-nosed form (see Plate 32) that looks very different from its crimson-spotted ancestor (see Plate 4) in the headwaters of the Darling River.

The main requirements of all species are relatively clear waters so that the spectacular colours of displaying males can be seen by prospective mates, and mats of submerged or floating plants for shelter and as an attachment site for the eggs. Relatively warm temperatures are essential for spawning, and in the far south the Goulburn rainbow may undergo dramatic population decreases during drought winters, when a combination of frost and low water levels reduce water temperatures below what it can survive except in the deepest and most sheltered refugia.

The closely related **blue-eyes** are mostly smaller than rainbowfishes, and some may be found in estuaries as well as in fresh waters, forming large schools in some places. Their spawning is also similar to the rainbowfishes with dramatic displays by the males, and though each female can only produce several eggs at a time, they may spawn over a very long season. The honey blue-eye (see Plate 4) is another endangered indigenous fish which is well protected on paper, with heavy fines for anyone who captures it from the wild, yet the primary threat to this species is rampant land clearing and housing development through much of its limited range.

Hardyheads are also small, slender schooling fishes, mostly marine but with a few living permanently in inland waters which may be extremely saline and exceptionally hot (for a fish habitat) in some cases. While some species are potentially endangered, others can be abundant in parts of their range and are a significant food source for waterbirds such as pelicans and cormorants when wet years swell their numbers. Despite their ecological importance and intermittent abundance, little is known of the breeding biology of many species, which does not help with conservation efforts.

Of the two species of **southern smelt**, one is widespread in the still and slow-flowing waters of south-eastern Australia, and breeds among submerged plants in fresh waters. The Tasmanian species is mainly found in coastal streams, spending part of the year in brackish waters but returning to fresh waters to spawn. Both are schooling fishes, and were probably much more important as both a minor predator and as prey in inland waters before European settlement, but they remain reasonably abundant at some times and places.

The **eel-tailed catfishes** are a very distinctive group with a fringe of fin around most of the rear part of the body, some of them looking like a compressed type of eel apart from the conspicuous whiskers (barbels) around the mouth. Most of them are tropical to subtropical, with two species extending south to south-western Australia and through the Murray–Darling. The barbels are sensitive organs of both taste and touch, well adapted to foraging for insects, crustaceans and molluscs and sometimes even smaller fishes in murkier waters, and they crawl readily with backs exposed in the shallows when feeding or the urge to migrate upstream through floodwaters takes them.

Little is known about breeding in most species, only the freshwater catfish of south-eastern Australia and the closely related freshwater cobbler of the south-west having been studied in any detail for aquaculture purposes, as both are good to eat. Males of these two species build a nest of pebbles or smooth gravel in which the eggs are laid, keeping them above any surrounding silt, but this may be abandoned if water levels drop too quickly before hatching. Other species that breed in sandy or stony places don't build a nest in the few cases where anything detailed is known of the breeding behavior.

The **fork-tailed catfishes** are a separate family that remains more closely tied to its marine origins, and are often found in estuaries as well as fresh waters. Already mentioned as mouth-brooders, these are usually solitary when adult but schools of young fishes are not uncommon. In the few cases where breeding localities are known these fishes also seem to favour coastal and near-coastal localities.

The bottom-dwelling **gobies** are closely allied to **gudgeons** (discussed earlier; some zoologists regard gudgeons as a less-specialised type of goby), the most obvious difference between them being the fused pelvic fins found in most gobies. These form a curved disc which can act as a suction cup, allowing gobies to cling to rock and other smooth surfaces with little energy expenditure, a particularly useful behavior for those which live in fast moving or turbulent waters.

Gobies are the largest family of marine fishes, and include the smallest vertebrates with adults of some species not even reaching the dimensions of a slender fingernail clipping, while other miniscule species live out their entire lives in less than two months. There are also many estuarine species that probably return to more saline waters for breeding, and a relatively small number that live out their entire lives in inland waters. As with gudgeons, gobies lay eggs on hard surfaces such as under rocks, with the parents guarding them until they hatch.

In Australia, the most extreme adaptations are found in the various desert gobies, tolerant of extremes of hardness and salinity, and temperatures from not that far above freezing to hot waters so depleted of oxygen that they must cling at the water's surface, keeping their heads above water to reach the air. At another extreme are cling-gobies, brightly striped fishes of pebbly, coastal streams with fairly constant temperatures and high oxygen levels, although like many other gobies found in fresh waters, these possibly migrate seasonally to the sea to breed.

Of all the gobies, the mangrove-dwelling mudskippers (see Plate 7) are perhaps the most intriguing, and have carried the intelligence, inquisitiveness and various other specialties of the family to new heights – literally. Able to breathe air and climb, with projecting eyes alert for danger from above, these active and territorial fishes spend more time above water than in it, and may drown if kept submerged for too long.

13

Amphibians and reptiles

Reptiles and amphibians are inextricably linked in most people's minds, though the physical differences between them are considerable, and they have very different ecological needs. These can be summarised very simply: amphibians have soft, scale-less skin with limited tolerance for drying out, and their soft-shelled eggs are usually laid in water or wet places. Frogs are the only representatives of this group in Australia, and because of their specialised needs the great majority of them are wetland animals for at least a part of their life cycle.

By contrast, the scaly skin of a reptile is designed to keep moisture in so desiccation is not a problem except in extreme conditions, so many reptiles live in arid places. Others have become secondarily associated with water, though the presence of open water is not always necessary to their well-being. The hard or leathery shell of a reptile egg also allows them to breed a long way from any water supply, though it also means the eggs would drown if laid in water, and to get around this problem some aquatic reptiles have become livebearers whose eggs develop into miniatures of the adult inside the mother.

Frogs – a disappearing act

All native amphibians in Australia, including those variously referred to as froglets, toads, toadlets and burrowing frogs are true frogs, even though the bodies of some may be squatter and more 'warty' looking than those of their more slender and stylish relatives. The only true toad in Australia is the toxic and unlovable cane toad (see Plate 19), discussed with other introduced vermin in Chapter 6.

Many frog species worldwide are in rapid decline, and at least eight in Australia have effectively disappeared from the wild in the past two decades, while others are rapidly declining in numbers and range. For example, the vividly yellow-striped corroboree frog is possibly down to below 50 calling males in the high country, well below the functional limits for a self-sustaining population, many populations of the green and golden bell frog and the closely related growling grass frog are in decline or have disappeared and researchers are not optimistic about their long-term future, while both gastric-brooding frogs have almost certainly become extinct not long after they were first discovered.

This has led to a proliferation of so-called frog-friendly ponds in backyard gardens, and publication of any number of guides to their construction, from pamphlets to booklets. Though some of these are so useless they make frog experts livid, at least one has been republished because it was favourably reviewed by an ignorant yet well-meaning media. Worse still, some of this literature encourages people to transfer frogs (usually in the form of tadpoles) from place to place, almost certainly contributing to the spread of the fungus implicated in the disappearance or ill-health of many species.

Chytridiomycosis has been linked to the disappearance of many native frogs, as well as amphibian declines and total disappearances worldwide. Although there may well be other factors involved in the extinctions, the most alarming scenario is that global warming has created ideal incubation temperatures for this disease even in upland areas where it was once absent.

Given the rapid rate at which it has taken hold in a wide range of climates, however, I suspect that diverse new and active strains have been spread worldwide by the aquarium industry. Even the most stringent import procedures won't detect diseases that captive bred fishes may carry if they don't affect the fishes themselves. These farms in places like Florida and Singapore also harbor wild frogs, which may carry any number of diseases our native species have no immunity to, without being affected themselves. Chytridiomycosis can now be identified in living frogs by a bath which detects fungal DNA, though there is still no cure in sight.

Stopping the spread of the fungus is difficult, perhaps impossible – it is already affecting frogs in relatively pristine rainforests and on the top of Mount Kosciusko, and waterbirds undoubtedly carry the plague with them on their far-flung wanderings. As I work in various wetlands, to avoid helping its spread I use a number of pairs of gumboots and waders which can be sterilised between field visits, scrubbing them clean of mud and drying them out in the sun for several days before storing. Handling frogs is the guaranteed fastest way to spread the problem, and if you need to do this for any reason use wetted disposable surgical gloves.

Increasing levels of salinity are also probably having an impact on frog populations, particularly in swamps, marshes and other wetlands which are rarely flushed by fresh water. Most frogs are not obviously affected by levels up to the equivalent of around 6% of sea water (4 ppt of salt), though there must already be some effect on their well-being as beyond that point numbers and diversity decline rapidly, and none seem to survive levels above 12% (8 ppt). Increasing aridification concentrates salts through increased evaporation, so global warming represents a direct threat to many of the frogs of southern Australia, less so in the north where the wet season flushes most wetlands out very thoroughly.

As the calls of most (male) frogs are distinctive and easily detected even if the animals can't be seen, declines in numbers or species diversity may be obvious, and though this in itself tells us little about the reasons for the decline it does make frogs extremely useful 'sentinels' for environmental deterioration. Whether this knowledge will be any help to the frogs themselves remains to be seen, because it won't achieve much unless it is acted upon.

Frog life cycles and habitats

As would be expected for a group of animals with a long evolutionary history on this continent, native frogs occupy many habitats and use them in many ways. The basics of their biology, are much the same as most frogs and toads worldwide, the main exceptions being various species that don't lay their eggs in water but in other moist environments, including the extinct gastric brooding frogs which incubated their eggs internally.

The young or tadpole stage of frogs (see Plate 29) is very different from the highly evolved adults, and is more similar to the primitive condition from which all amphibians evolved. They are more specialised than most primitive amphibians were, however, with a long gut coiled up in their bulging bellies for more efficient use of space. Their lips are often specially adapted to scrape or break up algae and other soft foods, although some may also be partly carnivorous.

The tadpole is a feeding machine, effectively a larval stage designed to put on weight as quickly as possible, converting this into an adult frog before the waters it lives in dry out and leave it stranded. At metamorphosis, the tail begins to be absorbed at more or less at the same time as small legs first appear, so the young frog is increasingly able to get around on its limbs. Usually, by the time the tail has become small enough to not drag on the ground behind it, the tadpole can leave the water for long periods of time.

Adult frogs feed primarily out of water, and the most common diet is insects in one form or another. Sometimes this may be a specific group such as termites in the case of some desert frogs, but more usually a wider range of species is taken, basically whatever can be caught readily in season. Food items vary in size according to what the frog can swallow, so the smallest species feed mostly on smaller insects, while those with larger mouths are more opportunistic and may even take young snakes and rats, or hard-shelled prey such as snails.

Apart from a food supply, the main requirement for frog survival is shelter, as many other wetland animals prey on adult frogs if given the chance. Some prefer vegetation, particularly tussocky plants, while others may hide under rocks or fallen timber. Larger balks of timber are excellent insulation from the hottest sun, but are also vulnerable to fire as the dry, dead tree trunks of many eucalypts will burn completely away to ash if they catch alight.

Male frogs call to attract a mate, and the calls of most species are distinct and easily recognised once you are familiar with them. This means it is possible to identify the frogs in any wet area by sound alone, though you will often have to listen in on a regular basis over a whole year as different species breed at different times. Where they call from also varies between species, some hiding among dense vegetation, in smaller pools of water near larger wetlands, or while floating in open water. They may call in groups, setting up a chorus triggered by the first in each group to call. This is partly to make sure females aren't attracted to the first caller alone, but probably also helps to distract predators who would find it easier to locate solitary callers.

Frog-friendly ponds

A frog-friendly garden with a pond is likely to be a healthy environment compared to one where herbicides and insecticides are used, but can unintentionally alter the gene pool of local frog populations, and not necessarily for the better. The first pair or handful of frogs arriving in a new and unpopulated pond are likely to breed much more successfully than those in nearby areas, where there may already be considerable competition as well as a range of predators, so a single pair of frogs have the chance to produce hundreds of offspring in an artificially protected environment.

Few gardens will sustain large populations of frogs, so most of the spill-over will have to return to nearby wetlands. The founders of the garden pond population are not necessarily representative of their species, and if they have only bred successfully because of the sheltered nature of that environment, a flood of their offspring will only add to competition in surrounding wetlands. Here they may even become genetically dominant through sheer weight of numbers, even if a higher proportion of them are taken by predators due to their sheltered upbringing, or fail to attract as many mates as their wild siblings.

Frogs escaping overcrowded suburban gardens are also more likely to have come in contact with alien diseases including chytrid fungus, and have the potential to act as vectors carrying other new diseases into wild populations. The absence of any kind of predator in the garden situation will make this problem worse, because in anything like a wild situation the vast majority of tadpoles or young frogs don't make it to adulthood. Diseased frogs and those that aren't quite up to scratch are usually among the first to be eliminated by predators.

This is why frogs produce the number of eggs they do – statistically speaking, only a very small number of their offspring are likely to survive, a good thing in terms of the long-term health of the population. To put it in ecological terms, in real wetlands the vast majority of tadpoles are *food*, not destined to become frogs, and the parents breed on a prolific scale to compensate for this.

The problem of single pairs of frogs breeding up on a grand scale in backyard ponds isn't necessarily only a matter of disproportionate injection of their genes into wild populations. We don't know that much about the relationship between different species of frogs, but it is already obvious that some species colonise garden ponds much more readily than others. Excessive buildup of species that thrive best in the relatively unnatural environment of a garden pond could easily have a negative effect on wild populations of other species nearby.

This also puts the debate about whether any fishes should be included in the pond into a different perspective. In the wild, huge numbers of tadpoles are consumed by diverse predatory waterbirds, by turtles, fishes, water spiders and mudeyes, and while they are still tiny even by backswimmers and stinging animals such as hydras. It may not seem fair to the frog enthusiast, but the process also helps to keep frog populations in balance with their local environment.

Let us put the 'problem' of fishes eating tadpoles into a wider perspective. First of all, many frogs in southern Australia breed in the cooler months when the majority of fish species are relatively inactive, and by the time the fishes are feeding more enthusiastically, the tadpoles are often large enough that they are not of interest as prey. I have had up to three of our local frogs breed successfully at one time in 5000 litre ponds with Goulburn River rainbowfish, purple-spotted gudgeons, and various carp gudgeons, all of which ignored the tadpoles present as these were already half the size of the fishes by the time they became active.

Even while small, many tadpoles are distasteful, and unless fishes are present in excessively large numbers and no other foods are available, each individual will usually sample only a few tadpoles before moving on to better-flavoured prey. Still other fishes such as pygmy perches, smaller rainbowfishes including threadfin and ornate rainbows, blue-eyes, hardyheads, dwarf galaxiids and most of the smaller gobies won't touch even the smallest tadpoles unless they are starving. Aquarium studies where tadpoles are the *only* food available to such fishes are not comparable to situations in the wild, where diverse invertebrates will also be available, and are favoured as food.

The main problem with fishes in frog ponds in that they are usually sourced from aquarium shops and have been bred by aquarists, often with little care as to the other fishes and potential diseases they have been in contact with. These may include piscine tuberculosis, chytrid fungi, diverse parasites and any number of other diseases, only a few of which will affect the frogs themselves, but a further danger is that frogs emigrating from garden ponds with disease-carrying fishes may export new diseases to nearby wetlands and other groups of animals.

Major groups of native frogs

Native frogs are classified in four separate families, two of which are only found in specialised habitats in the far north of Australia, and are relatively poorly known so that even the calls of some species have yet to be recorded. The Ranidae are an abundant family elsewhere, but the only native species is the wood frog which is still reasonably common on Cape York. The 15 or so species of Microhylidae are mostly localised in moist, forested, often mountainous areas of northern Queensland, with one species in the Northern Territory, and are not discussed here as their eggs are laid on land, developing into froglets away from wetlands.

Most Australian frogs belong in either the Myobatrachidae, a diverse family found only in the Australian region, or in the Hylidae. Members of the latter group are more recent arrivals from the north, but have been here long enough that Australian species form their own subfamily, different enough from the overseas representatives of the group that it has been suggested they should be placed in a new family called the Pelodryadidae.

Hylids include nearly all of the climbing treefrogs in this country, though there are also more specialised species living in arid areas, a niche also adopted by many myobatrachids which are mostly ground-living species with limited climbing ability. Most members of both families breed in water, but while the majority of hylids can climb out via twigs or overhanging foliage that reaches the water's surface without difficulty, myobatrachids need broader, shallowly angled surfaces such as branches laid over a pond's edge, or graded beaches, as neither adults nor metamorphosing tadpoles will be able to get out otherwise and will ultimately drown.

Both families have been in Australia for a very long time and have had abundant time to specialise for diverse habitats, so there are few worthwhile generalisations that can be made about them, and it is essential to research the life cycles of each and every species in a particular wetland or stream for management purposes. For example, the burrowing frogs of arid regions have developed this habit as a response to desert conditions, not because they are necessarily closely related. Only the major groups of both families that are associated with wetlands, including ephemeral or highly seasonal ones, are discussed in the brief accounts below.

Southern frogs (Myobatrachids) were evolving in Australia long before the hylids probably arrived, and are correspondingly diverse. Many of the smaller species are semi-terrestrial, often congregating in large breeding groups soon after rain fills ephemeral pools and grassy depressions. Although the majority of these don't have familiar common names they are often described as **froglets**. *Crinia* species are found all over Australia, in many closely related forms that may only be readily separated by their calls. This means that non-calling males and all females are almost impossible to identify in the field, though you can usually get a good idea of species from which part of the country they occur in.

The similar-looking *Geocrinia* species are mostly more southern in range, and are generally more seasonal breeders that respond rapidly to autumn or winter rains, with some breeding later in spring. In the first year we lived in the Otways, at least 50 mating pairs of the eastern common froglet appeared in a small, shallow and completely barren pool that surrounded our car within several hours of the first autumn downpour, crossing at least 50 metres of machinery-churned mud to reach it.

Toadlets of the genus *Pseudophryne* are mostly eastern and south-eastern in range, though there are several species in the west. Many of them are vividly coloured on the back, sides or underneath, presumably as a warning to predators that they are toxic. All of them are semi-terrestrial, breeding in marshy patches among forest or rocks, while the endangered corroboree frog is found above the tree line in wet, grassy areas and sphagnum bogs. *Uperoleia* toadlets are such small and cryptic animals that they don't have bona fide common names, just contrived ones bestowed on them in recent years. Primarily tropical to subtropical, many of the species live in arid regions or are from places with a marked difference between the wet and dry seasons, and many are burrowers.

The true **burrowing frogs** are a diverse group of animals which aren't necessarily closely related, and with varying degrees of specialisation. *Heleioporus* species are found mostly in the south-west of the continent, with one in the south-east, the common names of many species referring to their bizarre calls. These are large, blunt-featured creatures with cat-like eyes and broad feet with little sign of webbing, better adapted to burrowing than to water. They are found in moister woodlands rather than in truly arid areas, hiding their foamy egg masses in secluded places such as under vegetation from around late summer to autumn. As with *Neobatrachus* species which are also mostly western in range, most of these animals only venture to the surface to breed after rains.

Spadefoot toads are among the most specialised burrowers, with shovel-like feet and almost spherical bodies for retention of water in arid zones. Although these are true frogs, their common name refers to the warty appearance of the skin, and they are only likely to surface after the occasional rainfall to breed in temporary pools. Most specialised of all is the turtle frog which is no longer a wetland animal in any sense, feeding on termites and laying its eggs underground.

Barred frogs are handsome creatures with distinctive striped patterns on their hind legs, camouflaged in patterns and colours that make some of them look very leaf-like. These are largely terrestrial animals that live in wetter places in the forests of eastern Australia, often in gullies and along streams, where they breed in rock pools and backwaters. Barred frogs are disappearing from many parts of their range, suggesting a particular sensitivity to chytrid fungi.

Marsh and banjo frogs (see Plate 24) are as diverse as the habitats they are found in, including some species that are adapted to arid areas and can burrow fairly well, though most are from areas with more permanent waters. These are mostly medium-sized to fairly large frogs, and most of them lay their eggs in a conspicuous floating foam mass which probably provides some protection against predators. These are among the most prominent of frogs in semi-urban areas, and some such as the striped marsh frog may build up in large numbers around garden ponds, at the expense of other, smaller frogs.

Treefrogs (Hylids) fall into two natural groups, of which the true **treefrogs** are the most familiar, and although these are more diverse and abundant in warmer areas (especially towards the tropics), three species have reached as far south as Tasmania. All of them are good and often active climbers as well as jumpers, and they are usually found near streams and more permanent waters, especially in their inland range. Several species can also burrow, although they are nowhere near as specialised for this purpose as the myobatrachids discussed earlier.

While some treefrogs including the bell frogs are highly sensitive to chytrid fungi and are disappearing as a result, others are extremely adaptable and it is likely that some will spread further south with global warming. The eastern dwarf treefrog is already colonising parts of Melbourne well south of its native range, probably after accidental introduction among banana bunches, while the green treefrog (see Plate 12) is a familiar resident of less than pristine habitats such as toilet bowls in warmer climates.

Cyclorana is a group of treefrogs that have specialised for burrowing, and are found mainly in the northern half of Australia; some species show a marked resemblance to spadefoot toads, even down to the warty skin. These are mostly arid zone frogs, though some may be found in a wider range of habitats including coastal floodplains, woodlands and even monsoon forests, emerging from underground in the wet season, or after rains fall in drier places.

Freshwater turtles

Freshwater turtles have traditionally been called tortoises in Australia, a word better confined to purely terrestrial animals with very different habits and needs, none of which are found in this country. The word terrapin has also been used, meaning a type of semi-aquatic tortoise, but the term freshwater turtle is now widely accepted as the best description for our primarily aquatic shelled reptiles. This term also clearly separates them from marine turtles, all of which have well-developed flippers.

The only freshwater turtle with flippers is the **pig-nosed turtle**, originally described from New Guinea though it is also now known from northern Australia. Although not uncommon in some places once researchers realised it was there and started looking for it more closely, it appears to be a creature of larger rivers, often found in the tidal sections or in murky water. Both eggs and adults are eaten by humans in Papua New Guinea, but in northern Australia it appears to be fairly widely distributed and hopefully not under any serious threat, as there is little that could be done towards its conservation other than maintaining tidal rivers in undisturbed condition.

All other native freshwater turtles have legs, and some can travel considerable distances over land when the wetlands they are in dry up, though others will burrow into mud and aestivate for months through hotter, drier periods. Aestivating animals may also burrow into moist soil in shaded places, or be partly hidden under fallen timber where temperatures don't climb too high. They will also abandon places where water quality is seriously deteriorating, or where there is little in the way of food.

In the cooler parts of eastern Australia most turtles will hibernate during the cooler months when they are less active, and there is less food about. Although some will emerge from water to hibernate, many will stay underwater for days or even weeks at a time, possibly keeping up their oxygen supply from water being pumped through their cloaca. For some unknown reason, the two turtles of south-western Australia including the oblong turtle (see Plate 23) don't seem to hibernate, but remain more or less active throughout the year.

There are two main groups of freshwater turtles, the long-necked (snake-neck) types, and short-necks. The various **long-necked turtles** are found through most of the better-watered areas of the country, from southern Victoria and south-western Australia northwards along the coastal belts up into the tropics and into New Guinea, as well as throughout the Murray–Darling.

Short-necked turtles are a species-rich group found inland in the Cooper's Creek and Murray–Darling drainages (see Plate 32), but with a greater wealth of species around eastern and northern Australia where it seems almost every larger river system has its own distinct forms. Another group known as **snapping turtles** is largely tropical in range, with a few distinct forms found as far south as New South Wales. The strong jaws of these animals allow larger animals to feed on mussels, either by crushing the shell or by biting through where the two valves close together.

There are also several monotypic groups which are mostly river animals which seem fairly secure (and largely inaccessible) in their riverine habitats. The **western swamp turtle** is now only found in two ephemeral swamps in south-western Australia where it is protected legally as well as by fox-proof fencing, and by well-maintained fire breaks. Populations are also being increased by artificial hatching and feeding of the young, in a program that would serve well as a model for other endangered turtles, as well as endangered local populations of otherwise common species.

Nearly all freshwater turtles feed in water by preference, hunting smaller fishes and larger crustaceans such as shrimps and crayfish, and feeding on anything else they can catch, or even carrion in some cases. Some individuals within the same population may have their own specialist tastes, feeding selectively on particular animals, or may just be particularly good at catching them compared to their relatives. Where tadpoles are present these are often eaten in huge numbers, which is probably one of the reasons frogs that breed in more extensive swampy areas produce such large numbers of young.

Swimming and diving are also turtle defense mechanisms, and a disturbed animal can stay under water for long periods of time. Turtles need to bask out of water in the sun, usually when the heat is not too intense, and this may be important in keeping their shell firm and resistant to disease. Wild animals are extremely vigilant during basking, and are difficult to creep up on unawares.

Larger turtles were a favoured aboriginal food wherever they were found, and still are in some places, but are otherwise largely free of aquatic predators. Crocodiles are the only animal able to eat whole adult turtles, though larger ariid catfishes will swallow smaller adults as well as hatchlings, and eels may be a major predator in southern waters. All native species have scent glands which appear to be used for defense, producing a more or less offensive odour if they are disturbed or frightened. Some populations of snake-neck turtles are even known as 'Murray stinkers', for obvious reasons if you have ever been naïve enough to carry one around in a pocket.

The breeding requirements of the **common snake-neck** (see Plate 5) have already been discussed in Chapter 1, and in a general way these serve as a model for all other freshwater turtles, although there are variations in the types of soils, conditions and locations each species prefers for their buried nests. These vary not only between different species, but sometimes even between populations of a single species in different areas. Females will usually seek out a nesting area well above potential flood levels, and may travel considerable distances to find one, even climbing to well above the normal water level of the rivers they live in.

The eggs take months to hatch, and in some cases live young have been recorded in them nearly two years after the nest was dug. The nest is vulnerable to predators through all this time, of which the worst in contemporary Australia is now probably the introduced fox, although native predators such as goannas will also take a minor toll. As the adults live for many decades, this can give a false sense that a population is secure, yet few eggs may be hatching to replace the ageing adults. On the other hand, an apparent lack of young may be equally deceptive in another way, as they are extremely shy and are rarely seen.

Lizards

Most lizards are primarily terrestrial animals, but quite a few species of **skinks** (Scincidae) live in the vicinity of streams and wetlands, and some of these are active swimmers which will even feed opportunistically on small fishes trapped in shallow waters. To the north these are most often *Carlia* species, but in eastern Australia most of those referred to as water skinks are *Eulamprus* species, subspecies and local forms.

Although mostly found in and around creeks and wetlands, and excellent swimmers which take to water readily, they lay their eggs on higher and drier ground wherever possible. Our local water skinks (a variant on *E. tympanum*) multiplied prodigiously after we set up our wetland nursery in the late 1980s, and although they spend most of their time in and around the nursery ponds, their eggs are only found in cracked clay and mounds of rock dust.

Among the larger lizards, only one of the dragon family (Agamidae) is regularly aquatic. The eastern water dragon (see Plate 5) spends most of its time in trees overhanging the water or foraging near the water's edge, but will leap into water and remain submerged for long periods if disturbed. Left to itself it rarely if ever goes swimming, and although it isn't often seen far from water it is likely that the central place of water in its life is primarily as a refuge.

In tropical Australia, several of the smaller monitor lizards (Varanidae, also known as goannas) are found in and around streams or mangroves in tidal areas most of the time. These feed above water on anything from insects to frogs while they are still young, graduating to larger prey including fishes and even carrion once they reach an adult size. One monitor at least is threatened by the spreading range of the cane toad; Merten's water monitor can swallow toads large enough to kill it, and anecdotal evidence suggests dramatic declines in population as the cane toad invasion has swept westward across the tropical north.

Crocodiles

There is an abundant literature on these top-level predators, most of which is irrelevant to most people establishing or managing a wetland, and it need not be repeated here. Essentially, if you have crocodiles in your wetland it is either very large, and you are probably managing it as part of a public reserve of some kind, or you have a problem. Even the relatively harmless freshwater crocodile needs an adequate supply of fairly large fishes, and this will have a flow-on effect on the ecology of the wetland as a whole. This may not be a problem for passive management, however, as the crocodile is likely to move on once its food supply has been thinned enough. In smaller wetlands, crocodiles (especially the dangerous saltwater species) may be seasonal visitors, particularly during the wet season, and will usually be removed to other waters by fisheries officials on request.

Snakes

Few Australian snakes are aquatic, and the majority feed on lizards as their primary food, unlike most snakes in other lands where small mammals are more abundant so they are the main prey. Though almost none could be regarded as true wetland animals, native snakes swim well and quite a few indigenous species include frogs as a significant part of their diet so they are often found near water.

The most aquatic species are two species of **file snakes**, rough-skinned, live-bearing, tropical creatures which are graceful if sluggish in water, but barely animate out of it. The combination of extreme clumsiness out of their preferred element, and an exclusive diet of fish means that they are found only in large wetlands with significant fish populations, and where they can swim to new wetlands during floods. A file snake in a smaller wetland is in a dead end from which it can't escape, and which is unlikely to provide it with an adequate food supply.

These are animals of the great floodplains of the far north, and they remain abundant even though they have long been a favoured food among aboriginal peoples. The **water python** shares the same extensive environment, but is a more versatile species feeding opportunistically on a wide range of foods both in and out of water. When wetlands are dry, water pythons will actively seek rats in cracks in the ground, while in the wet they become much more aquatic and hunt smaller birds or their eggs.

The snakes encountered in most wetlands are often just on their way through, though some will stay to hunt frogs for months or even years if these are abundant. In southern Australia visitors of this kind are mostly venomous, but the dangers they pose are greatly exaggerated (see later in this chapter). Tiger snakes, brown snakes and black snakes all feed on frogs when they can, but will usually move out of the area before long, especially if disturbed too many times.

There are exceptions; a **copperhead** lived in my wetland nursery for over a decade, lolling about on the paths on sunny days even when my wife or I were walking past regularly, but vanishing if disturbed by the footsteps of an unfamiliar visitor. In years when frogs were particularly abundant, small batches of her young would be found around our swimming dam, dispersing after a few weeks as they grew larger and exhausted the supply of smaller frogs.

Copperheads are active in much colder weather than other temperate area snakes, and I have seen them swimming in search of prey on sunlit days at the end of winter, emerging to warm up in the sun every few minutes. **Red-bellied black snakes** will also feed in and near water regularly, and may even catch small fishes when these are abundant, though these are usually only a minor part of their diet. **Keelbacks** are the only native snake which not only feeds on frogs, but can also eat smaller cane toads and their tadpoles with apparent impunity. Other wetland species include the so-called water snakes, and the much less frequently seen mangrove snakes.

Like all potentially dangerous creatures, venomous snakes generate a bizarre wake of myths and odd beliefs, few of which stand up to scrutiny. In the context of wetlands, it is worth putting to rest the myth that snakes can't open their mouths

How dangerous are snakes?

Frogs live in wetlands, and many snakes including some of our most venomous species eat frogs, so anyone who lives or works near wetlands is likely to encounter them sooner or later. Although several people die of snakebite each year in Australia, in most cases the snake has bitten in self-defence, because these humans were trying to kill it. Statistically speaking, you are a thousand times more likely to die in a car accident on the way to a hospital, than from an *unprovoked* bite!

In reality, even the most supposedly aggressive species such as taipans (which aren't frog eaters, so they don't usually hunt near water) and tiger snakes will retreat if given the slightest chance, unless they feel trapped or are being directly threatened. I have only ever been struck at by one snake in over 40 years working around water, a tiger accidentally trapped against a pond wall which raced towards me in an obvious panic, giving up and turning sideways into the bush as soon as I had sprinted 20 metres away.

Even the reputedly most aggressive snakes are distinctly timid. A large tiger snake trapped in a fruit net on our property suffocated and died just as I cut its head free, its body set into an awkward-looking knot. It was only 15 minutes after I had carried the rigid body into the bush by its middle that I realised it should have been limp, with *rigor mortis* not developing until sometime later. Sure enough, the snake was gone when I returned to check; it had been literally scared stiff of me as I handled it!

Striking snakes don't necessarily inject venom – they may give a 'dry' bite just as a warning. The venoms is made up of complex biochemicals which take energy to manufacture, so they are reluctant to use it except in self-defence or on prey. Although the chances of a serious snakebite are small, there are simple, common sense precautions anyone working or living around wetlands should take. Apart from keeping a sharp lookout among plants near the water's edge, move relatively slowly and walk fairly heavily so the vibrations of your footsteps carry. Frog-hunting snakes are most active around dusk, or around dawn on hot days, and are also most difficult to see at these times.

Long, baggy pants are the safest wear even in the hottest weather, and gumboots or waders are even better if temperatures are moderate and you are working near water. Always carry a pressure bandage with you, at all times, and know how to apply it. Though potentially deadly, the venom of native snakes doesn't cause massive local damage at the bite site, unlike that of dangerous snakes in some other countries. The venom in a bitten limb wrapped firmly with a pressure bandage should take hours to move into the rest of the body, giving plenty of time to get to a hospital.

Accurate identification of the species will make it possible to apply the right antivenin and in the right quantity, to minimise side-effects of the cure. If you can't safely kill the snake to take it with you to the hospital, leave the bite site unwashed, as a swab sample of the venom should be enough for identification. See Cogger (2000) for a full account of how to deal with snakebite.

while swimming, for fear of drowning. Snakes in water are usually there to hunt frogs, and if they can't open their mouths this begs the question of how they catch their prey (hint: it isn't by hypnosis).

It is also common to hear tales of people being pursued by a serpent for great distances. The known speed of the fastest snakes in the world is barely half that of a running human, and their stamina is poor. Independent witnesses to the chase are likely to have been treated to the sight of a terrified human running a kilometre or more before daring to look back, while the exhausted 'pursuer' lay hyperventilating at least 950 metres behind.

14

Birds and mammals

Birds have already been discussed as wildcards in wetlands, some as efficient predators, others arriving in large groups and by their sudden presence changing the entire ecology of a place. They have also featured in many other different roles in this book, and although I have cautioned against evaluating the success of a wetland purely in terms of the birds it draws, and highlighted the problems caused by excessive populations, they are also an important part of the beauty and fascination of wetlands.

I never grow tired of watching black swans taking off, or a V-formation of pelicans coming in to land like a great, gawky flock of pterodactyls. And though I like to think I have a reasonably orderly and rational mind, if a crake crosses my path I know it is a good omen for the day to come. The very nature of these three examples also tells us that the term 'waterbird' does not define a tidy and closely related animal group – waterbirds are as ecologically diverse as the fishes.

Swans are herbivores which feed in the open shallows or over grassy flats near water, able to migrate great distances with their powerful flight, mating in pairs for life and defending mounded nests of vegetation built out away from shore. Crakes are cryptic creatures lurking and nesting among dense vegetation and feeding on invertebrates, shoots and perhaps seeds, running rather than flying from cover to cover. Pelicans don't care much for vegetated waters as these interfere with their communal fishing tactics, and nest in groups on open islands or peninsulas preferably far from shore, working hard to take off but able to glide effortlessly for hours once they are airborne.

Flight paths

Despite their great diversity, one thing most larger waterbirds have in common is the need to be able to fly in and out of wetlands easily. This is part of the reason more open waters attract a greater range of species and proportionately larger numbers of birds, though finite food resources are also an important component as has been discussed in Chapter 9. Trees planted too close to water were also discussed - the smaller the dam, the more of a confining effect these have so that any waterbird foolish enough to land will feel trapped in the bottom of a deep and windless pit.

This increases the difficulty of taking off, especially as many larger waterbirds are fairly clumsy fliers and need a considerable distance to achieve lift-off even with a stiff breeze in their faces to help with the aerodynamics. A few islands large enough for trees scattered around an extensive and otherwise open wetland are not a problem, and with enough water between them and the shore will even be used as a roost by some species of waterbird. Widely spaced stands of smaller trees such as paperbarks standing in or beside the water may also be used as communal breeding sites by egrets, cormorants and ibises.

As well as being rather clumsy on take-off, many waterbirds are also not very manoeuvrable as they descend to land, and in urban areas electricity cables strung across flight paths are dangerous to swans and pelicans in particular. Brightly reflecting or vividly coloured markers spaced every few metres along the cables will help give advance warning to incoming birds so they can change their course a little and avoid the danger, and although unsightly at least these look better than the numerous corpses I have seen scattered near wires in some wetlands.

Major groups of birds associated with wetlands

Defining a wetland bird is a difficult task even on a group-by-group or individual basis, although there are many obvious contenders that no-one would dispute. The brief descriptions below of major groups associated with water are not intended to be comprehensive, but will give a good indication of the range and diversity of the species most likely to be found in or near water for a significant part of their lives.

Swans, ducks and geese are almost the definitive waterbirds, certainly the first to come to most peoples' minds when thinking about wetlands even though some species are mostly terrestrial in their habits and some take to the water reluctantly. Most native species are migratory to a greater or lesser degree, sometimes moving on in search of less ephemeral wetlands, and sometimes just apparently through an innate restlessness in their non-breeding season.

Black swans (see Plate 28) have already been used as an example of a large, herbivorous bird, and although they may aggregate in considerable numbers in more permanent waters during times of drought, they are aggressive and pairs require a reasonable area to successfully raise their young. This means that in drought years most pairs will breed later than the more successful ones with established territories, sometimes as late as spring once rains have filled other wetlands in the area, and there is enough vegetation growing to make a nest and feed the breeders. Swans will also breed in dams that have a large enough shallow area and clear water that allows abundant plant growth, in other words where livestock have been fenced out.

Ducks and geese are not always easily separated, and some of the birds we usually call ducks are effectively miniature geese in terms of their behavior and ecology. Of the true **geese** the magpie goose is now usually regarded as a northern species but it formerly bred much further in southern Australia until it was hunted and poisoned out, or abandoned areas which had been too extensively drained. Although it nests in large flocks in wetlands on hollow sedges, it will fly rather than try to escape from potential predators by swimming.

This grazing bird is now reappearing sporadically in parts of its southern range in small numbers, most often in pairs as the males will fight dramatically even when feeding together in large flocks. By contrast, the Cape Barren goose is primarily a terrestrial grazer which thrives best on offshore islands, migrating over considerable distances at times, and though it is also found near wetlands at times it will rarely swim and breeds on dry land.

Shelducks are the largest of the goose-like ducks, and may spend much of their time grazing on land though they also feed on various types of algae and even floating fairy fern. The mountain duck will often aggregate in large flocks to moult near water after their breeding season. Although they may breed near saltmarshes and saline areas, their young must have access to fresh water once they leave the nest in tree hollows and limbs, sometimes falling as much as 20 metres before they reach the ground.

Pygmy-geese are the smallest geese and are mainly confined to far northern areas now. Unlike most native ducks and geese they seem reluctant to travel any great distances, though they ranged much further south in the past until habitat destruction and possibly the blanketing effects of the introduced water hyacinth affected many of their favoured habitats. They may mate for life, and feed upon a range of primarily vegetable foods from seeds of waterlilies and aquatic grasses as well as submerged plants such as pondweeds.

Whistling-ducks (also sometimes called tree-ducks) are closely related to geese and swans and are named for their whistling flight, forming huge flocks in northern parts of Australia. The water whistling-duck is a truly aquatic animal that feeds on aquatic plants, dives readily, and is rarely seen away from the water. By contrast the plumed whistling-duck is a largely terrestrial grazer and rarely swims; it is also far more migratory in habits and smaller groups can be found as far south as the Murray River at times. Like geese and swans they may mate for life, and nest on the ground at the onset of the wet season, not necessarily near water.

Most Australian ducks are **dabbling ducks**, so-called because they are not able to dive deeply, feeding instead by dabbling among plants and in shallow water, and sieving water through their beaks or effectively standing on their heads to reach the bottom. The serrations inside the beak are almost as effective as teeth in grinding and tearing, and can grip painfully if a duck seizes your finger! The best-known species include the common black duck as well as teals and shovellers, and these are the birds most frequently associated with open water in urban areas. If present in unnaturally large numbers because they are being fed by well-meaning urbanites, they can cause considerable damage to aquatic plants which make up the bulk of the diet of some species, and can also undermine and collapse the shoreline with their constant probing along the water's edge.

The black duck will appear almost anywhere there is water, and will even nest in long grass at the side of an apparently barren dam if cattle are kept away for the breeding season, though it prefers tree hollows if these are available. Teal are more fussy about their nesting sites, reportedly preferring rocky islands, but they will lay eggs in any secluded spot at the water's edge. Females of most dabbling ducks raise their young without help from the male, except in the case of the chestnut teal (see Plate 22) which seems to form a more lasting pair bond.

The **wood duck** (see Plate 27) is another of the more successful species in adapting to changes since European settlement, and is often seen near farm dams in large numbers as well as in urban areas. It will graze in groups near water, and a pair can raise up to around 20 young if a hollow tree is available near a favoured dam with abundant grazing around it, but they are not particularly water-loving and will also breed in trees well away from water.

The **musk duck** looks surprisingly reptilian when swimming, a heavy-looking creature sunk deep in the water and moving with powerful thrusts of its legs to the sides. It is among the most aquatic ducks and has difficulty moving around on land, flying mainly by night. Not surprisingly for such an able diver the diet includes a range of larger invertebrates, mostly insects but also crayfish, mussels, and sometimes even frogs. The male is aggressive during courtship and leaves the female to raise young herself, in a nest built among dense reeds; she may also lay eggs in the nests of other ducks after the fashion of a cuckoo.

The **Australian pelican** (see Plate 8) has already been briefly mentioned, and is a scavenger as well as a fishing bird. Though pelicans will hunt in groups (sometimes in association with schools of cormorants) circling a school of fish, they will also happily forage in tips and are perhaps the best known of all predatory waterbirds as they are not ashamed to beg. Far from a typical predatory waterbird, the pelican is primarily designed for a life at sea or in open waters and can catch and swallow some impressively sized prey.

Herons and egrets vary considerably in size and appearance, but this is a single natural family of which the egrets are the most distinctive group simply because they are white, and develop striking plumes during the breeding season. Long-legged and long-beaked stalking predators of shallow waters, all species will feed on fishes, frogs, larger invertebrates and some even on young birds and small mammals. Most species forage in relatively open and gently sloping areas, though some are more opportunistic and will hunt even among dense vegetation if there is enough food present to justify the effort.

All species prefer to nest in trees, some in pairs or small groups only, but egrets especially prefer communal breeding areas which they may share with ibis and even cormorants. Some species are much more specialised than most of their family, for example the Nankeen night heron which only appears at dusk, and will hunt through the night. **Bitterns** are even more specialised, highly secretive and well camouflaged among the reed beds they nest in. The brown bittern is the most familiar species though even it is rarely seen except by people looking specifically for it, though its powerful booming call is not uncommon in more extensive marshes.

The herons and egrets are elegant fliers with their heads tucked back into the body, and legs trailing behind, while (among other differences) **storks** and **cranes** fly with their necks outstretched, though they are not otherwise very similar or closely related. The jabiru is the only native stork, once more widespread but now largely confined to the north as it is not particularly migratory. It is a clumsy hunter, sweeping its beak through shallow waters and seizing fish, snakes, frogs and any other larger animals it bumps up against. Like the brolga, the most widespread of Australia's two native cranes, the jabiru mates for life though it nests in trees near fresh water while the brolga makes a nest of vegetation in shallow water or on islands.

Unlike all the other long-beaked, long-legged water birds, brolgas include a considerable amount of vegetable matter in their diet; in the north mainly tubers of bulgurru sedge, while in southern Australia they often work their way across stubble fields, feeding on fallen grain and any mouse or cricket they find along the way. At night they retreat to shallow wetlands where they will stand at discrete distances, making them less accessible to predators.

Cormorants (see Plate 8) and **darters** are clumsy fliers, especially once their wings are waterlogged, so they need to spend considerable time drying out after diving for their prey. They are excellent swimmers and divers, however, and can stay under water for a few minutes at a time in search of fishes, tadpoles and sometimes crayfish. All species nest in colonies though darters are usually in smaller groups, some species at ground level on islands well out from shore, but more usually in shrubby vegetation and small trees well above the ground.

Grebes are even more specialised for a swimming and diving existence than darters, with a stiffened fringe around each toe instead of webbing. They can swim long distances submerged in pursuit of fishes and larger aquatic invertebrates, and are rarely seen out of water – even their low-set, floating nests are usually hidden among hollow-leaved sedges well out from shore. The little grebe is the most familiar species and will breed in suitable farm dams, forming small flocks at other times. Its young look remarkably like furry, striped toy turtles, and are good swimmers from an early age.

Several **raptors** also hunt around larger and more open wetlands, but the osprey and the white-bellied sea-eagle are mostly coastal and estuarine birds and need not be discussed here. The marsh harrier lives much of its life in and around open, sedge wetlands but also other grassy places near water where it can see to hunt, nesting in smaller trees near the water but not necessarily dependent on this habitat.

Unlike many other waterbirds which often congregate and nest in large groups, the long-billed **kingfishers** are usually solitary hunters although pairs will generally be found within whistling distance of each other. These ambush predators are found in a wide range of habitats where the water is usually still and clear enough for their prey to be visible including estuaries, mangroves and larger streams. They take small to medium-sized fishes, frogs, crayfish and crabs from a perch well above the surface, killing with a beating against a branch rather than a stab wound, and their nest building is as specialised as their hunting techniques with a tunnel being made into a creek bank or a termite nest depending on species.

Not all kingfishers are regularly associated with wetlands, and some such as kookaburras are distinctly terrestrial in their habits. Those that hunt along streams and in mangroves are mostly blue, and for me even a brief glimpse of the spectacular and jewel-like azure kingfisher glittering in sunlight is the highlight of a day spent near water. Kingfishers are never abundant and they avoid places where there is too much visible human activity, so this is one group of waterbirds that will benefit from a reasonably tall screen of trees or shrubs near the water's edge, preferably 10 or more metres wide.

Ibises (see Plate 27) and **spoonbills** are closely related, the obvious differences being in the ways their very differently shaped, elongated bills are used to feed. Both groups are long-legged waders, making nests in shrubs, tangled lignum and sedge beds, sometimes in very large, mixed colonies. The down-curved ibis bill is used more as a probe, locating smaller animals including insect larvae under water, but straw-necked and sacred ibis will also feed over wet pastures in mixed flocks, searching for crickets, worms and grubs.

Spoonbills feed with a side-to-side action of the beak through water, seizing mostly very small prey from tiny fishes to insects and crustaceans, so they need to feed repeatedly and often for long periods of time to satisfy their needs. The beak is snapped shut on anything that comes in contact with the inside of the spoon, but they will also sometimes forage out of water. Spoonbills travel in much smaller groups than ibis, often with both species together, and a single individual may linger in a particular dam if there is enough food to support it for some time.

Rails and **crakes** are the smallest members of the rail family, cryptic birds that hide among vegetation and are usually only briefly seen as they dash across an open area. However, they may be quite common even if not often seen, and a patient observer lying low on the fringes of any densely vegetated marsh may get to observe individuals feeding for some time, or in breeding season even the occasional fight.

Crakes are very small and extremely shy birds with relatively stumpy beaks, while those of the larger rails are sharper looking. Their diet is a combination of invertebrates and vegetable matter including seeds, and they are presumably fairly flexible in what they eat as they can be found across a range of wetland types including saltmarshes with abundant shrubby cover. All species are reasonable swimmers and some may dive to escape attention, making nests of aquatic vegetation just above water level, often with a shallow ramp leading down.

Native hens are effectively larger rails in habit as some of their common names suggest, but are less shy and are usually reasonably easily observed except when living among tall sedges. These are mostly indigenous birds though one species is found into South-East Asia and Polynesia, and several of their relatives are found across much wider areas. The **Eurasian coot** (see Plate 8) and **dusky moorhen** are true aquatic birds that include a significant amount of vegetation in their diet, supplementing this with any invertebrates too slow to escape.

Moorhens prefer to stay closer to shelter and can stay underwater for minutes if frightened, while coots may spend much of their day on more open waters in large flocks, presumably as a partial protection against raptors. The **purple swamphen** (see Plate 26) is largely vegetarian and less aquatic, and will happily pick its way over suburban lawns – and through wetland nurseries! All three species build low, rounded nests of aquatic vegetation, sometimes in surprisingly exposed situations, but the well-camouflaged eggs are difficult to see even from a metre or two away.

The **lotusbird** is distantly related to shorebirds such as plovers, and is the only reasonably common and widespread jacana in Australia. Extremely long toes allow it to walk across waterlily and lotus leaves, turning the edges of these to feed on various invertebrates below as well as the seeds of the aquatic plants it is always found among. Even the floating nests are made from aquatic plants, and the patterning of the black-scribbled eggs matches the colour of the blackening, dying stalks.

The **clamorous reed-warbler** is found throughout Australia and invariably weaves its nests among the stems of the common reed (see Plate 26) where its eggs are protected from the approach of most predators by the water below. Although rarely seen for more than a brief moment at a time, its melodic repetitions of a few simple syllables are one of the defining sounds of reedy wetlands in spring.

Shorebirds

Migratory shorebirds are not considered in any detail here as many of them are only seasonal visitors to Australia, not even necessarily nesting here. The wide open wetlands and muddy shorelines, often including seashores, these birds frequent are nearly all protected and managed under the Ramsar convention, an international treaty covering the seasonal habitats of migratory wetland birds which range over as much as half the planet each year. Conveniently, these also happen to be the most important habitats for many of the shorebirds that do breed in Australia. Small groups and individuals of some shorebird species may appear along the drying shores of larger dams and relatively plant-free wetlands, but they usually move on to join larger flocks within a day or two.

There are exceptions – an odd (literally) individual of Latham's **snipe** spent its southern hemisphere season wandering around our ornamental garden of camellias, apparently nostalgic for the gardens of Japan, emitting occasional alarming shrieks but showing no interest in the wetland nursery 60 metres away. This species is fairly easily seen as it flushes into the air when disturbed, but other species such as the painted snipe are so secretive almost nothing is known about them, and there is no way to even predict where they will appear from time to time.

Many shorebirds feed over mudflats, probing for any small invertebrates they can find from molluscs to worms and burrowing crustaceans, but others will forage in drier places and include insects and even seeds as a significant part of their diet. Still others such as red-necked **avocet** and **stilts** (see Plate 28) are long-legged so they can feed in deeper areas than other species, and a greater part of their diet is made up of crustaceans such as brine shrimp and also insect larvae. They breed in colonies which may number in the thousands for some species, near the inland saltpans and open wetlands where they feed.

Various **plovers, dotterels** and **lapwings** are also often associated with open wetlands, though most of them are more generally associated with coastal landscapes. Many species are only occasional vagrants in Australia, and even those that breed here regularly often nest far from the nearest water. The nest is an open scrape in the ground, sometimes lined with shells, feathers or aquatic vegetation, and usually both parents raise and defend the young.

Mammals

Compared to birds, aquatic mammals are mere bit players in the scheme of things. Although they can move around they don't usually travel great distances, and will usually have a well-defined home range. In turn, that means their populations must be closely tied in to the available food resources throughout the year, so family groups are the largest gatherings likely to be found together, and then only during the breeding season.

That doesn't make mammals of less significance than birds, though it is their conservation that is of more concern than any strong ecological influence they bring to bear, as they can easily be driven from a section of river or wetland never to return. Most species associated with wetlands are fringe dwellers found in places near more permanent waters, but by and large living a more terrestrial lifestyle. Only the two most aquatic native mammals are briefly included here for the sake of completeness, and because I am more than passing fond of them both.

The **platypus** (see Plate 16) is among the most fascinating of all mammals, though its history and evolutionary origins have been repeated so often that there should be no reason to describe it here; there is also a regularly revised book (currently in its fourth edition) dedicated to it alone. Feeding mostly around dawn and dusk its main diet is a diverse array of macroinvertebrates, most of them on the small side but some up to the size of a small freshwater crayfish, and it will also take tadpoles and possibly small fishes if given the opportunity.

This is a truly unique and iconic animal, and people come from all over the world for the chance to see them in the wild. I have been lucky enough to take out tour groups occasionally in canoes on a nearby lake, and the experience is magical, from walking in through the dark among treeferns with constellations of glow-worms below that are literally as bright as the stars, to the first sighting as the mists lift and the light brightens. It is a sight that has probably not changed much in tens of millions of years.

The **water rat** is far more adaptable yet less glamorous despite attempts to rename it with various attractive but localised aboriginal names such as rakali. Almost otter-like compared to other native rats, with a tidy toothbrush of bristles along each side of the nose, it is found from subalpine areas to the sea. When the water rat family that lived around a suburban lake where I used to catch yabbies as a child moved away as their prey declined in numbers, I was pleased to find them re-established among rock piles under a jetty in nearby Port Phillip Bay.

Although water rats are much more common than the platypus they are seen less often, the main sign of their presence usually being little middens of the remains of meals on rocks and logs near the water. An adaptable carnivore, the water rat will take crayfish, mussels, surprisingly large fishes if they can be caught in a shallow pool, and even whole lamb chops intended to be cut up for yabby bait. Another dawn and dusk feeder, like the platypus it also retreats into tunnels it has made in inconspicuous places along the shoreline.

Glossary

aestivation: a state of dormancy during a hot or dry period.

amphipod: aquatic crustaceans, the ancestors of terrestrial leafhoppers which look much the same.

animal: in the broad zoological sense used in this book, any bird, invertebrate, fish, reptile or amphibian.

annual: a plant that lives only long enough to set seed for the next generation, sometimes up to a year but usually less.

artesian waters: underground waters brought up from some depth, used in drier areas as a source of stock and irrigation water, but usually very hard.

biota: collective term for all living things including plants, animals, bacteria and fungi.

bottleneck: a restriction in genetic variability, common in small or isolated groups of plants or animals, and causing 'inbreeding'.

brackish: a mix of fresh and saline waters, in no particular proportions.

capillary: a fine tube or narrow space, through which liquids are drawn upwards through their own surface tension.

catchment: the land surface from which the water to a dam or wetland is collected as runoff.

cloaca: the single opening through which some animals (including reptiles and frogs) mate and excrete.

copepod: tiny, free-swimming, shrimp-like crustaceans with no common name, often abundant and an important food for larger animals in both fresh and saline waters.

crustacean: major invertebrate group with an exterior skeleton, distantly related to insects, and including shrimps, water fleas, copepods, crayfish and leafhoppers.

cyanobacteria: blue-green 'algae'.

dampland: an area where the watertable rises to very close to the soil surface during some times of the year.

electrofishing: temporarily stunning fish with a powerful, localised electric current from a generator in a backpack, or on a boat.

endemic: found only in a particular area.

ephemeral: in wetlands, a body of water which regularly dries out for a part of the year.

estuarine: found in estuaries.

estuary: the section near a river or creek mouth where sea and fresh waters mix.

exotic: introduced from another area, usually another country in the sense used here, but also a 'native' that doesn't belong.

feral: wild population of an animal descended from escaped domestic or captive stocks.

fry: newly hatched or very young fish.

genetic drift: an inbreeding effect which happens when a very small number of plants or animals colonise a wetland, and reproduce to form a closely-related population which may be somewhat different (drift away from) the norm for that species.

germination: in seeds, sprouting or starting into growth.

groundwater: water pooled or flowing underground.

hardness: a measure of the quantity of calcium, magnesium and other salts present in water, other than sodium salts (see also **salinity**).

heathland: seasonal wetland, often on sandy, peaty soil.

herbivorous: feeding on plants.

hydrology: the study of water as it flows over- or underground, and associated changes to its properties.

indigenous: native to a particular area, in other words not introduced from elsewhere.

interstitial: living in between the particles of soil or sand.

invertebrate: animal without a bony internal skeleton or backbone.

larvae: immature stages of an insect, often very different in appearance from the adult stage.

limnology: the study of inland (most often fresh) waters and their biota.

lunette: crescent-shaped mound of sediments, usually built up on the fringe of a lake by wind, during a period of drought.

mangroves: trees adapted to living in areas where water levels rise and fall with sea tides, and salinity can vary from almost pure sea water to fresh water at different times.

milt: fish sperm.

monotypic: with only a single species in a genus or family.

mudeye: carnivorous larval stage of a dragonfly.

native: used loosely in this book to mean any animal or plant which originates in some part of Australia, and sometimes in the sense that it belongs where it is found (i.e. indigenous).

omnivorous: feeding upon both plant and animal foods.

peat: plant matter which has only partly decomposed because of the absence of nitrogen and some other nutrients, which is very absorbent so it acts as a water store in swampy soils.

phytoplankton: plant plankton, particularly single-celled algae including diatoms.

plankton: freely drifting animals, often with little or no control over their direction of drift.

pH: a scale used to compare acidity or alkalinity, centred around a neutral point of 7.0. pH readings below 7 are increasingly acid as numbers go down, above 7 are increasingly alkaline as numbers go up.

ppm: abbreviation for parts per million.

ppt: abbreviation for parts per thousand.

raptors: hawks, falcons, ospreys and related predatory birds.

riparian: found along streams, usually used to describe vegetation.

salinity: a measure of the quantity of sodium salts present in water (see also **hardness**).

salmonid: a fish of the salmon family, in Australia, trout or occasionally salmon.

soft: refers to water with low salinity and relatively few minerals.

seed shrimp: crustaceans resembling a tiny, swimming clam, often abundant but not apparently favoured as a food by many other animals.

spawning: fertilising eggs, in the case of aquatic animals often (but not always) outside the body in water.

sumpland: a dampland where the watertable rises up to and sometimes a little above the soil surface at some times of the year.

terrestrial: growing or living on land, rather than in water or very wet places.

tributary: a smaller stream feeding into a larger river system.

twitcher: a serious bird watcher, though not necessarily lacking a sense of humour.

vector: an animal capable of carrying a disease affecting other species, even if it is not necessarily affected by the disease itself.

water flea: cladoceran crustaceans, so-called for their clumsy, hooping swimming style.

water table: the upper surface of underground water, which may rise or fall with the seasons. This can be located by digging down to the point where a hole fills with water to a certain level – the water 'table'.

Recommended reading

This was originally intended to be a comprehensive bibliography, but after a quarter of a century of compilation, during the course of which some really excellent general references have been published, it has been whittled down to those books (and a fair few articles) which are good starting points for readers wanting to learn about any particular field.

Wherever possible I have cited recent books with a complete and relatively current bibliography for their subject area, so that more specialised and regional guides can be located through these. Any significant guides published more recently than the bibliographies of these major texts are included by way of an update.

Internet research and making contacts

Although a number of useful websites are included with the various entries below, the internet is also a useful starting point for locating recent and detailed information on the biology of many wetland animals. Search for the most recent scientific name you have available for a particular species, but before following up any sources that appear, check through the first few dozen of these to see if there have been any recent changes to the scientific name, or you will probably miss much of the more recent research available.

For example, a search for the brownback crab under *Holthuisana transversa*, the name it has been known under for decades, turns up close to 600 entries at the time of writing and confirms that this is the correct spelling rather than *Holthuisiana* as in Williams (see below, Williams 1980). Many of these articles are preoccupied with how a crab, from a primarily marine group, regulates its salt balance in fresh waters, and as it spends much of its time out of water, various aspects of its breathing mechanisms.

Some sources mention the breeding habits of this crab though there are few technical papers on this aspect of its biology. Most importantly, a few taxonomic papers reveal that it is now regarded as unrelated to other (non-Australian) freshwater crabs in *Holthuisana*, and is correctly *Austrothelphusa transversa*. This means that an ever-increasing number of more recent papers and articles will be citing this name alone, and a search under the new name reveals around 270 hits, most of which will not turn up in a search for *Holthuisana*.

Refining the search by adding the word 'reproduction' to the scientific name of any species being researched will bring up most technical papers on this area. Separate searches on 'breeding' or 'spawning' will sometimes turn up amateur or hobbyist accounts, which often include useful information that is not easy to winkle out of more technical papers. As this crab survives drought in arid areas in sealed burrows, for this particular species the keywords 'burrow' or 'drought' will bring up more specific articles on this aspect of its biology.

Technical papers give you two further sources of information, including a reference section that will often lead you to earlier articles which may include quite a bit on the broader aspects of the biology of any species, and which is not repeated in later papers because it is now regarded as common knowledge. It will also give you the authors' names and where they were working at the time of publication; a routine search will tell you whether they are still in the same place, or where they have moved.

These researchers are likely to know everyone working in the same field, and will usually be pleased to know that someone other than other specialists is interested in their work. In many cases they will also be happy to email you copies of their recent papers, and suggest who else you could contact for other information you are looking for – assuming that all the answers are known, of course.

If you don't like using the internet, contacting the appropriate curator at your State museum will lead you to much of the same information less directly. In the case of the brownback crab, the curator of invertebrates is unlikely to know more than the basics of its biology, but will certainly know who is working on it in the same State, and possibly elsewhere. Keep in mind that while curators and researchers will usually be happy to help, they have work of their own to get on with and will not do your research for you. Don't waste their time with basic questions; it is more productive for everyone if you don't contact them until you have done enough research to ask intelligent questions!

General

Bennett A, Backhouse G & Clark T (1995) *People and Nature Conservation: Perspectives on Private Land Use and Endangered Species Recovery.* Transactions of the Royal Zoological Society of New South Wales, Sydney.

Brearly A (2005) *Ernest Hodgkin's Swanland: Estuaries and Coastal lagoons of South-western Australia.* University of Western Australia Press, Perth.

Chorus I & Bartram J (Eds) (1999) *Toxic Cyanobacteria in Water: A Guide to their Public Health Consequences, Monitoring and Management.* E & FN Spoon/ Routledge, London.

De Dekker P (1986) What happened to the Australian aquatic biota 18,000 years ago? In: *Limnology in Australia*. (Eds P De Dekker & WD Williams) pp.487–96. CSIRO, Melbourne.

Lange K & Partners (1974) *A Preliminary Study of the Ecology and Hydrology of Lake Colac, Victoria*. The City and Shire of Colac, with Environmental Resources of Australia.

McComb AJ & Lake PS (1990) *Australian Wetlands*. Angus and Robertson, Sydney.

Norris RH, Liston P, Davies N, Coysh J, Dyer F, Linke S, Prosser I & Young W (2001) *Snapshot of the Murray-Darling Basin River Condition*. Murray–Darling Basin Commission, Canberra.

Romanowski N (2007) *Sustainable Freshwater Aquaculture: The Complete Guide from Backyard to Investor*. University of New South Wales Press, Sydney.

Saintilan N (Ed) (2009) *Australian Saltmarsh Ecology*. CSIRO Publishing, Melbourne.

Serena M (Ed.) (1994) *Reintroduction Biology of Australian and New Zealand Fauna*. Surrey Beatty & Sons, Chipping Norton.

Sexton M (2003) *Silent Flood: Australia's Salinity Crisis*. Australian Broadcasting Commission, Sydney.

Streever WJ (Ed.) (1997) Wetland Rehabilitation in Australia. *Wetlands Ecology and Management* special issue **5**(1), 1–97.

Turner L, Tracey D, Tilden J & Dennison WC (2004) *Where River Meets Sea: Exploring Australia's Estuaries*. Cooperative Research Centre for Coastal Zone, Estuary and Waterway Management, Brisbane.

Williams WD (Ed.) (1998) *Wetlands in a Dry Land: Understanding for Management*. Environment Australia, Canberra.

Young W (Ed.) (2001) *Rivers as Ecological Systems: The Murray-Darling Basin*. Murray–Darling Basin Commission, Canberra.

Invertebrates

Readers with a particular interest in insects will find *The Waterbug Book: A Guide to the Freshwater Macroinvertebrates of Temperate Australia* (Gooderham J & Tsyrlin E, 2002, CSIRO Publishing, Melbourne) particularly useful, though this also covers many other common wetland invertebrates briefly.

The best general guide to a wider range of animals is *Australian Freshwater Life: The Invertebrates of Australian Inland Waters* (Williams WD, my original edition dates back to 1980 and was published by Macmillan). This is widely available in libraries, though the names used are sometimes dated, and readers needing accurate names for some groups will need to refer to more recent and specialised works.

There are as many detailed identification guides available as there are major groups of invertebrates, the widest array of which were published by the

Murray–Darling Basin Commission (now the Murray–Darling Basin Authority), and these can be sourced through their website at www.mdba.gov.au. Some other useful references covering various invertebrates, and issues associated with their conservation are included below.

Anderson NM & Weir TA (2004) *Australian Water Bugs: Their Biology and Identification*. CSIRO Publishing, Melbourne.

Davis J & Christidis F (1997) *A Guide to Wetland Invertebrates of Southwestern Australia*. W. A. Museum, Perth.

Ponder W & Lunney D (1999) *The Other 99%: The Conservation and Biodiversity of Invertebrates*. Transactions of the Royal Zoological Society of New South Wales, Sydney.

Suthers IM & Rissik D (Eds) (2009) *Plankton: A Guide to their Ecology and Monitoring for Water Quality*. CSIRO Publishing, Melbourne.

Theischinger G & Hawking J (2006) *The Complete Field Guide to Dragonflies of Australia*. CSIRO Publishing, Melbourne.

Fishes

Many regional books on fish species are dated in parts. Allen, Midgley & Allen (2002) is the most accessible recent source of (brief) updates on names and species, and includes all of the significant regional and national fish faunas in its recommended reading list. A more detailed summary of sources for south-eastern Australia can be found in McDowall (1996).

Allen GR, Midgley SH & Allen M (2002) *Field Guide to the Freshwater Fishes of Australia*. Western Australian Museum, Perth.

Cowx IG & Welcomme RL (Eds) (1998) *Rehabilitation of Rivers for Fish*. Fishing News Books, Cambridge, UK. (Useful for general principles, but with a particular emphasis on the habitat needs of carps and salmonids.)

McDowall RM (1990) *New Zealand Freshwater Fishes: A Natural History and Guide* (2nd edition). Heinemann Reed, Auckland.

McDowall RM (Ed.) (1996) *Freshwater Fishes of South-Eastern Australia*. Reed Books, Sydney.

Odeh M (Ed.) (1999) *Innovations in Fish Passage Technology*. American Fisheries Society, Bethesda, Maryland.

Pusey B, Kennard M & Arthington A (2004) *Freshwater Fishes of North-Eastern Australia*. Centre for Riverine Landscapes, Griffith University, Nathan, Queensland.

Romanowski N (2004) Notes on Dwarf galaxias *Galaxiella pusilla*. *Fishes of Sahul* **18**(4), 80–86.

Schweid R (2009) *Eel*. Reaktion Books, London.

Wager R & Jackson P (1993) *The Action Plan for Australian Freshwater Fishes*. Australian Nature Conservation Agency, Canberra. This can be downloaded from www.environment.gov.au/biodiversity/action/fish.

The Australia New Guinea Fishes Association (ANGFA) publishes a quarterly journal in high-quality colour. Although this is primarily an aquarists association with a particular interest in rainbowfishes, the journal also regularly publishes articles by researchers on various other fish groups and sometimes macroinvertebrates, and there are also local groups which meet regularly for talks and field trips. See www.angfa.org.au for contact details.

Reptiles and amphibians

Anstis M (2002) *Tadpoles of South-eastern Australia: A Guide with Keys*. Reed New Holland, Sydney.

Cann J (1998) *Australian Freshwater Turtles*. Beaumont Publishing, Singapore.

Cogger HG (2000) *Reptiles and Amphibians of Australia* (6th edition). Reed New Holland, Sydney.

Collins JP, Crump ML & Lovejoy TE (2008) *Extinction in Our Times: The Global Amphibian Decline*. Oxford University Press, New York.

Goldingay R & Osborne W (Eds) (2009) Ecology and conservation of Australian bell frogs. Special edition of *The Australian Zoologist* **34**(3), 325–460.

Shine R (1998) *Australian Snakes: A Natural History*. Reed New Holland, Sydney.

Thompson MB (1983) Populations of the Murray River tortoise: the effect of egg predation by the Red fox. *Australian Wildlife Resources* **10**, 363–371.

Tyler MJ (1999) *Australian Frogs: A Natural History*. Reed New Holland, Sydney.

Tyler MJ & Doughty P (2009) *Field Guide to Frogs of Western Australia*. Western Australian Museum, Perth.

Tyler MJ & Knight F (2009) *Field Guide to the Frogs of Australia*. CSIRO Publishing, Melbourne.

Webb G & Manolis C (2002) *Australian Crocodiles: A Natural History* (2nd edition). Reed New Holland, Sydney.

Wilson S (2005) *A Field Guide to Reptiles of Queensland*. Reed New Holland, Sydney.

There are also numerous reptile and amphibian societies around Australia, which are loosely affiliated as the Australasian Affiliation of Herpetological Societies. A complete, fairly recent list of websites and addresses for these can be found in *What Snake is That? Introducing Australian Snakes* (Swan G & Wilson S, 2008, Reed New Holland), which also includes a good general summary of the biology and ecology of most of the more common snakes.

The site maintained by the Amphibian Research Centre at www.frogs.org.au includes contacts for local frog groups Australia wide, and general regional guides to all Australian frogs with a useful selection of pattern variations, rather than just single photos to illustrate entire species.

Mammals

Grant T (2007) *Platypus* (4th edition). CSIRO Publishing, Melbourne.

Menkhorst P & Knight F (2001) *A Field Guide to the Mammals of Australia*. Oxford University Press, Melbourne.

van Dyk, S & Strahan, R (Eds) (2008). *The mammals of Australia*. Reed New Holland, Sydney.

Birds

Bird lovers are blessed with an abundant literature, usually well illustrated with photography and original artworks, as well as several identification guides which are all included here without comment as to their relative merits as everyone has their own ideas on this subject!

There are also many regional books which are potentially of much wider application and usefulness than their titles may suggest – to cite just one, *Birds of French Island Wetlands* (Quinn D & Lacey G, 1999, Spectrum Publications, Melbourne).

Barker RD & Vestjens WJM (1989) *The Food of Australian Birds* (2 volumes). CSIRO Publishing, Melbourne.

Frith HJ (1982) *Waterfowl in Australia*. Angus & Robertson, Sydney. Although long superseded by other and more specialised books, this is still worthwhile introductory reading for ducks, geese and swans.

Geering A, Agnew L & Harding S (2007) *Shorebirds of Australia*. CSIRO Publishing, Melbourne.

Hollands D (1999) *Kingfishers and Kookaburras: Jewels of the Australian Bush*. Reed New Holland, Sydney.

Kingsford RT, Thomas RF & Wong PS (1997) *Significant Wetlands for Waterbirds in the Murray-Darling Basin*. National Parks and Wildlife Service, Sydney.

McKilligan N (2005) *Herons, Egrets & Bitterns: Their Biology and Conservation in Australia*. CSIRO Publishing, Melbourne.

Morcombe M (2003) *Field Guide to Australian Birds* (2nd edition). Steve Parish Publishing, Brisbane.

Pizzey G & Knight F (2007) *Field Guide to the Birds of Australia* (8th edition). HarperCollins Publishers, Sydney.

Simpson K & Day N (2004) *Field Guide to the Birds of Australia* (7th edition). Penguin Books, Melbourne.

Slater P, Slater P & Slater R (2003) *The Slater Field Guide to the Birds of Australia*. New Holland, Sydney.

The National Photographic Index of Australian Wildlife (1985) *The Waterbirds of Australia*. Angus & Robertson, Sydney.

The two major, nationwide birders clubs in Australia are the best starting point for information on everything from groups with specialised interests, to tours, useful literature, etc. Contact these through their websites, The Bird Observers Club of Australia at www.birdobservers.org.au, and Birds Australia (Royal Australian Ornithological Union) at www.birdsaustralia.com.au.

Plants

For information on weeds and their control, www.weeds.org.au is the single most useful site and includes links to many other sites, as well as online reporting of new arrivals or outbreaks, and current information on biological control. *Waterplants in Australia* is a useful general guide to many of the most widespread and well-established species already in Australia.

Cowie ID, Short PS & Osterkamp Madsen M (2000) *Floodplain Flora: A Flora of the Coastal Floodplains of the Northern Territory, Australia*. Australian Biological Resources Study, Canberra.

Duke N (2006) *Australia's Mangroves: The Authoritative Guide to Australia's Mangrove Plants*. University of Queensland, Brisbane.

Entwisle TJ, Sonneman JA & Lewis SH (1997) *Freshwater Algae in Australia: A Guide to Conspicuous Genera*. Sainty & Associates, Sydney.

Romanowski N (1998) *Aquatic and Wetland Plants: A Field Guide for Non-Tropical Australia*. University of New South Wales Press, Sydney.

Romanowski N (2009) *Planting Wetlands and Dams: A Practical Guide to Wetland Design, Construction and Propagation* (2nd edition). CSIRO Publishing, Melbourne.

Sainty GR & Jacobs SWL (2003) *Waterplants in Australia: A Field Guide* (4th edition). Sainty and Associates, Sydney.

Index of common and scientific names

Blind cave eel – *Ophisternon candidum*

Blind gudgeon – *Milyeringa veritas*

Blown grass – *Lachnagrostis filiformis*

Blue-eyes – fishes of the family Pseudomugilidae

Brine shrimp – the introduced *Artemia salina* (there are also native species)

Broad-leaved paperbark – *Melaleuca quinquenervia*

Brolga – *Grus rubicundus*

Brook char – *Salvelinus fontinalis*

Brown bittern – *Botaurus poiciloptilus*

Brown treefrog – *Litoria ewingii*

Brown trout – *Salmo trutta*

Brownback crab – *Austrothelphusa transversa*

Bulguru – *Eleocharis dulcis*

Bungwahl – *Blechnum indicum*

Burrowing frogs – mostly *Heleioporus* and *Neobatrachus* species

Buttongrass – *Gymnoschoenus sphaerocephalus*

Canadian pondweed – *Elodea canadensis*

Cane toad – *Bufo marinus*

Cape Barren goose – *Cereopsis novaehollandiae*

Cape waterlily – *Nymphaea caerulea*

Cardinalfishes – the mainly marine family *Apogonidae*

Carp – the introduced *Cyprinus carpio*

Carp gudgeons – various *Hypseleotris* species

Cat-tail – the introduced *Typha latifolia*

Cattle egret – *Ardeola ibis*

Cherabin – *Macrobrachium rosenbergii*

Chestnut teal – *Anas castanea*

Chironomid larvae – young of non-biting midges in the family Chironomidae

Chytrid fungus – *Batrachochytrium dendrobatidis*

Cichlids – introduced fishes of the family Cichlidae (pronounced sick-lids)

Climbing galaxias – *Galaxias brevipinnis*

Cling-gobies – *Sicyopterus* and *Stiphodon* species

Clamorous reed-warbler – *Acrocephalus stentoreus*

Climbing maidenhair – *Lygodium japonicum*

Clover – various *Trifolium* species, all introduced

Coarse clubrush – *Isolepis prolifera*

Cods – some fishes of the family Perchichthyidae

Common amphipod – various species of *Pseudomoera*

Common clubrush – *Isolepis inudata*

Common jollytail – *Galaxias maculatus*

Common reed – *Phragmites australis*

Common snakeneck turtle – *Chelodina longicollis*

Congolli – *Pseudaphritis urvillii*

Copepod – tiny crustaceans including species of *Boeckella, Calamoecia*, etc.

Copperhead snakes – *Austrelaps* species

Cordrushes – plants of the family Restionaceae

Cormorants – several species of *Phalacrocorax*

Corroboree frog – *Pseudophryne corroboree*

Crakes – mostly *Porzana* species

Crimson-spotted rainbowfish – *Melanotaenia duboulayi*

Cumbungi – either of the two native species of *Typha, T. orientalis* and *T. domingensis*

Darter – *Anhinga melanogaster*

Desert gobies – several species of *Chlamydogobius*

Drain sedge – *Cyperus eragrostis*

Dusky moorhen – *Gallinulla tenebrosa*

Eacham rainbowfish – *Melanotaenia eachamensis*

Eastern little galaxias – *Galaxiella pusilla*

Eastern common froglet – *Crinia signifera*

Eastern dwarf treefrog – *Litoria fallax*

Eastern water dragon – *Physignathus lesueurii*

Eel-tailed catfishes – family Plotosidae, mostly species of *Neosilurus, Porochilus* and *Tandanus*

Eelgrasses – several *Vallisneria* species

Egrets – mostly *Egretta* species

Epizootic Ulcerative Syndrome – *Aphanomyces invadans*

Estuarine crab – *Amarinus lacustris*

Eurasian coot – *Fulica atra*

Fairy fern – *Azolla* species

Fairy shrimp – various *Branchinella* species

Feral pig – *Sus scrofa*

Fiery skimmer – *Orthetrum villosovittatum*

File snakes – *Acrochordus* species

Fishing spiders – species of *Dolomedes*

Floodplain water ribbon – *Triglochin dubia*

Fork-tailed catfishes – family Ariidae, mainly *Arius* species

Freshwater catfish – *Tandanus tandanus*

Freshwater cobbler – *Tandanus bostocki*

Freshwater crab – *Austrothelphusa transversa*

Freshwater crocodile – *Crocodylus johnstoni*

Freshwater prawns – various species of *Macrobrachium*

Freshwater shrimps – *Paratya* and *Caridina* species

Frogbit – *Hydrocharis dubia*

Froglets – *Crinia, Geocrinia* and *Paracrinia* species

Galaxiids – fishes of the family Galaxiidae

Gambusia – *Gambusia holbrooki*

Gastric brooding frogs – two species of *Rheobatrachus*

Giant fern – *Angiopteris evecta*

Glass shrimp – *Paratya australiensis*

Glassfishes – the family Ambassidae, mostly (but not all) species of *Ambassis*

Gobies – fishes of the family Gobiidae

Golden perch – *Macquaria ambigua*

Goldfish – *Carassius auratus*

Goulburn rainbowfish – the most southerly form of the Murray rainbow, *Melanotaenia fluviatilis*

Green and golden bell frog – *Litoria aurea*

Green treefrog – *Litoria caerulea*

Green waterspider – *Dolomedes* species (these may also come in other colour forms including blue)

Growling grass frog – *Litoria raniformis*

Grunters – diverse fishes of the family Terapontidae

Gudgeons – fishes of the family Eleotridae

Guppy – the introduced fish *Poecilia reticulata*

Hardyheads – fishes of the family Atherinidae, mostly *Craterocephalus* species

Herons – mostly *Ardea* species

Honey blue-eye – *Pseudomugil mellis*

Hydra – sea anemone relatives of genus *Hydra*

Hymenachne – *Hymenachne amplexicaulis*

Jabiru – *Ephippiorhynchus asiaticus*

Keelback snake – *Tropidonophis mairii*

Keelbacked water fleas – *Daphnia carinata* and relatives

Lampreys – fish-like animals in the families Geotriidae and Mordaciidae

Land yabby – various species of *Engaeus*

Lantana – Lantana species, most commonly *L. camara*

Lapwings – two species of *Vanellus*

Latham's snipe – *Gallinago hardwickii*

Little grebe – *Tachybaptus novaehollandiae*

Little pied cormorant – *Phalacrocorax melanoleucos*

Long-necked turtles – *Chelodina* species

Lotusbird – *Irediparra gallinacea*

Macquarie turtle – *Emydura macquarii*

Magpie goose – *Anseranas semipalmata*

Mallard – *Anas platyrhynchos*

Maned duck – *Chenonetta jubata*

Mangrove jack – *Lutjanus argentimaculatus*

Marine clubrush – *Bolboschoenus caldwellii*

Marron – *Cherax tenuimanus*

Marsh frogs – various species of *Limnodynastes*

Marsh harrier – *Circus aeruginosus*

Marsh yabby – various species of *Geocharax*

Maundia – *Maundia triglochinoides*

Merten's water monitor – *Varanus mertensi*

Mexican waterlily – *Nymphaea mexicana*

Mimosa – *Mimosa pigra*

Mosquitoes – biting flies of the family Culicidae

Mountain duck – *Tadorna tadornoides*

Mouth almighty – *Glossamia aprion*

Mudeye – a dragonfly larva, of any of dozens of genera

Mudskippers – *Periophthalmus* species

Murray cod – *Maccullochella peelii*

Murray River spiny crayfish – *Euastacus armatus*

Murray River rainbowfish – *Melanotaenia fluviatilis*

Musk duck – *Biziura lobata*

Nankeen night heron – *Nycticorax caledonicus*

Narrowleaf water ribbon – *Triglochin lineare*

Native brine shrimp – *Parartemia* species

Native hens – several *Gallinula* species

Nightfish – *Bostockia porosa*

Nurseryfish – *Kurtus gulliveri*

Orange-bellied parrot – *Neophema chrysogaster*

Oriental weatherloach – *Misgurnus* species (in Australia probably *M. anguillicaudatus*)

Ornate rainbowfish – *Rhadinocentris ornatus*

Osprey – *Pandion haliaetus*

Pacific heron – *Ardea pacifica*

Painted snipe – *Rostratula australis*

Pampas grass – *Cortaderia* species, usually *C. selloana*

Paperbarks – *Melaleuca* species

Para grass – *Urochloa mutica*

Parrotfeather – *Myriophyllum aquaticum*

Pedder galaxias – *Galaxias pedderensis*

Peruvian water-primrose – *Ludwigia peruviana*

Phantom midges – family Chaoboridae

Pig-nosed turtle – *Carettochelys insculpta*

Plague minnow – the introduced *Gambusia holbrooki*

Platy – the introduced fish *Xiphophorus maculatus*

Platypus – *Ornithorhynchus anatinus*

Plovers – in the context of inland wetlands, mainly *Pluvialis* species

Plumed whistling-duck – *Dendrocygna eytoni*

Pobblebonk frog – *Limnodynastes dumerilii*

Pondweeds – *Potamogeton* species

Purple-spotted gudgeon – *Mogurnda adspersa*

Purple swamphen – *Porphyrio porphyrio*

Pygmy-geese – two species of *Nettapus*

Pygmy perches – *Nannoperca, Nannatherina* and *Edelia* species

Pygmy water ribbon – *Triglochin alcockiae*

Quacking froglet – *Crinia georgiana*

Queensland lungfish – *Neoceratodus forsteri*

Rails – in wetlands, mainly *Rallus* species

Rainbow nardoo – *Marsilea mutica*

Rainbow trout – *Oncorhynchus mykiss*

Rainbowfishes – fishes of the family Melanotaeniidae

Red saltpan algae – *Dunaliella salina*

Red water-milfoil – *Myriophyllum verrucosum*

Red-bellied black snake – *Pseudechis porphyriacus*

Red-eared slider – *Trachemys scripta*

Redclaw – *Cherax quadricarinatus*

Redfin perch – *Perca fluviatilis*

Red-finned blue-eye – *Scaturiginichthys vermeilipinnis*

Red-necked avocet – *Recurvirostra novaehollandiae*

Reed sweetgrass – *Glyceria maxima*

River lily – *Crinum pedunculatum*

River pandanus – *Pandanus aquaticus*

River redgum – *Eucalyptus camaldulensis*

Robust water-milfoil – *Myriophyllum papillosum*

Rushes – *Juncus* species

Sacred ibis – *Threskiornis aethiopica*

Sacred lotus – *Nelumbo nucifera*

Salamanderfish – *Lepidogalaxias salamandroides*

Salmon catfish – *Arius* species

Salt paperbark – *Melaleuca halmatutorum*

Saltwater crocodile – *Crocodylus porosus*

Salvinia – *Salvinia molesta*

Saratoga – one of two native fish species of *Scleropages*

Seven-spot archerfish – *Toxotes chatareus*

Sharp clubrush – *Schoenoplectus pungens*

Shelducks – *Tadorna* species

Sharp rush – *Juncus acutus*

Shield shrimps – *Lepidurus apus* and *Triops australiensis*

Short-finned eel – *Anguilla australis*

Short-necked turtles – *Emydura* species

Shovellers – two species of *Anas*

Skinks – lizards of the family Scincidae

Silverstripe mudskipper – *Periophthalmus argentilineatus*

Silver perch – *Bidyanus bidyanus*

Slender cumbungi – *Typha domingensis*

Small-fruited water-mat – *Lepilaena bilocularis*

Smooth crayfishes – diverse species of *Cherax*

Snapping turtles – *Elseya* species.

Soft treefern – *Dicksonia antarctica*

Sooty grunter – *Hephaestus fuliginosus*

South-western laceplant – *Aponogeton hexatepalus*

Southern freshwater prawn – *Macrobrachium australiense*

Southern smelt – two fishes of the genus *Retropinna*

Spadefoot toads – *Notaden* species

Spangled perch – *Leiopotherapon unicolor*

Spikerushes – *Eleocharis* species

Spiny crayfishes – *Euastacus* species

Spoonbills – two species of *Platalea*

Spotted livebearer – *Phalloceros caudimaculatus*

Spotted minnow – *Galaxias truttaceus*

Starwort – various *Callitriche* species, most commonly two which have been introduced

Straw-necked ibis – *Threskiornis spinicollis*

Striped grunter – *Amniataba percoides*

Striped marsh frog – *Limnodynastes peronii*

Swamp fern – *Stenochlaena palustris*

Swamp lily – *Ottelia ovalifolia*

Swamp she-oak – *Casuarina glauca*

Swordtail – the introduced fish *Xiphophorus helleri*

Tall flatsedge – *Cyperus exaltatus*

Tasmanian spiny crayfish – *Astacopsis gouldii*

Tasmanian whitebait – *Lovettia sealii*

Teals – several *Anas* species

Teatree – *Leptospermum* species

Threadfin rainbowfish – *Iriatherina werneri*

Tiger snake – *Notechis scutatus*

Tilapia – a generic name for a group of cichlid fishes, used in Australia for *Oreochromis mossambicus*

Toadlets – *Pseudophryne* and *Uperoleia* species

Treefrogs – *Litoria* species

Turtle frog – *Myobatrachus gouldii*

Twigrushes – *Baumea* species

Variable groundsel – *Senecio pinnatifolius*

Water boatmen – bugs of the family Corixidae

Water buffalo – *Bubalus bubalus*

Water couch – *Paspalum distichum*

Water ferns – *Blechnum* species

Water fleas – tiny crustaceans including species of *Daphnia*, *Moina*, etc.

Water hawthorn – *Aponogeton distachyos*

Water-holding frog – *Cyclorana platycephala*

Water hyacinth – *Eichhornia crassipes*

Water lettuce – *Pistia stratioides*

Water needles – various species of bug including *Ranatra*

Water python – *Liasis mackloti*

Water rat – *Hydromys chrysogaster*

Water ribbons – *Triglochin* species; the most abundant and widespread of these are the two variants of *T. procera*

Water scorpions – bugs of the genus *Laccotrephes*

Water shield – *Brasenia schreberi*

Water skinks – some *Eulamprus* species and forms

Water snakes – *Stegonotus* species

Water thyme – *Hydrilla verticillata*

Water whistling-duck – *Dendrocygna arcuata*

Waterbutton – *Cotula coronopifolia*

Watercress – *Nasturtium officinale*

Waterlilies – *Nymphaea* species

Western swamp turtle – *Pseudemydura umbrina*

Whirligig beetles – family Gyrinidae

Whistling-ducks – two species of *Dendrocygna*

White-bellied sea-eagle – *Haliaeetus leucogaster*

White-faced heron – *Ardea novaehollandiae*

White spot disease – *Ichthyophthirius multifilus*

Widgeon grass – several species of *Ruppia*

Willow herbs – various species of *Epilobium*

Willows – diverse *Salix* species and hybrids

Wolf spiders – family Lycosidae

Wood duck – *Chenonetta jubata*

Wood frog – *Rana daemeli*

Yabby – *Cherax destructor*

Yarra spiny crayfish – *Euastacus yarraensis*

Index

www.ingramcontent.com/pod-product-compliance
Lightning Source LLC
Chambersburg PA
CBHW041129280526
45792CB00013B/2366